BARR · DIAPORTHALES

The Diaporthales in North America

with Emphasis on *Gnomonia* and its Segregates

MARGARET E. BARR

MYCOLOGIA MEMOIR NO. 7

Published for
The New York Botanical Garden
In Collaboration with
The Mycological Society of America

J. CRAMER PUBLISHER
P.O.BOX 48,
3306 LEHRE GERMANY
1978

Author's address:

Department of Botany,
University of Massachusetts,
Amherst, Massachusetts 01003,
U. S. A.

LIBRARY OF CONGRESS CATALOG CARD NUMBER 78-12369
Printed in Germany
November 14th, 1978
by Strauss & Cramer GmbH., D-6945 Hirschberg 2
ISBN 3-7682-1189-4

CONTENTS

INTRODUCTION

This investigation had its genesis after I had accumulated a number of collections belonging to *Gnomonia* sensu lato, and subsequently had difficulties in identifying them. My intention at the time was to clarify the generic concept of *Gnomonia* and related taxa and to determine the limitations of species. As the study progressed, recognition of salient taxonomic characteristics of genera centered around *Gnomonia* presented a pattern which appeared to be applicable to other genera within the Diaporthales. As a result, the study has been extended to provide a rearrangement of the genera of Diaporthales, a brief description of each North American genus, notes on the relationships of extralimital genera and species, and detailed taxonomic studies on the species of *Gnomonia* and its segregates in temperate North America.

Acknowledgements

I wish to thank the National Science Foundation for financial support: Grant G 19534, held jointly with Howard E. Bigelow, resulted in many collections and stimulated my interest in this group of fungi, while GB 7996 enabled me to expand studies on the Diaporthales. The generosity of the curators of herbaria cited in the text in loaning collections in their keeping was especially valuable and is greatly appreciated. Cooperation of many mycologists has permitted the incorporation of valuable data. I acknowledge the comradeship, criticism, and support of Howard E. Bigelow in this endeavor. For advice in nomenclatural problems and for identification of host plants, I thank Mr. H.E. Ahles of the University of Massachusetts Herbarium.

Historical Survey

Fries (**1823**) knew a number of the species now placed in the Diaporthales and he arranged them in five of the sections of the composite genus *Sphaeria*, in at least nine tribes. When the genus was subdivided in subsequent years, these species were assigned to various genera in the family Sphaeriaceae. Fries (**1825**) erected *Hercospora*, and **1849** he segregated the genera *Endothia*, *Hypospila*, and *Valsa*. *Prosthecium* was erected by Fresenius (**1852**); *Pleuroceras* by Riess (**1854**), and *Mazzantia* by Montagne (**1856**). Cesati and de Notaris (**1863**) created the genera *Gnomonia, Mamiania,* and *Pseudovalsa.* L.-R. and C. Tulasne (**1863**) recognized *Cryptospora, Melanconis,* and *Melogramma*; Auerswald (**1869**) proposed *Cryptoderis.* Several genera were added or expanded by Nitschke (**1867**). Fuckel (**1870**) arranged diaporthaceous genera mainly in the Sphaeriacei of the Pyrenomycetes; the majority of nonstromatic species were grouped under the Ceratostomeae, while stromatic forms were distributed among the Melanconideae, Valseae, Dothideae, and Melogrammeae. Karsten (**1873**), in the family Sphaeriaceae, included *Pseudovalsa, Melanconis, Hercospora,* and *Cryptospora* in subfamily Melanconideae, *Diaporthe, Gnomonia,* and *Hypospila* in Diaportheae, *Valsa* in Valseae, and *Linospora* and *Ceuthocarpon* in Bertieae. Saccardo (**1882, 1883**) utilized ascospore septation and pigmentation for classification of the Ascomycetes. Under the family Sphaeriaceae, members of the Diaporthales are found in the various spore sections, excluding only the Phaeosporae, Hyalodictyae and Phaeodictyae; *Mazzantia* was inserted in the family Dothideaceae.

Winter (**1887**) divided his order Pyrenomycetes into recognizable families. In suborder Sphaeriaceae, the nonstromatic genera were arranged in the families Ceratostomeae, Clypeosphaerieae, and Gnomonieae, stromatic genera in the families Valseae, Melanconideae, and Melogrammeae. Again, *Mazzantia* was placed in the suborder Dothideaceae. Ellis and Everhart (**1892**) and Schroeter (**1893-1908**) utilized Winter's classification.

Lindau in Engler and Prantl (**1897**) recognized a number of genera which earlier had been given only subgeneric status. The families were similar to those of Winter, and diaporthaceous genera were arranged in the family Dothideaceae of Dothideales, and in the Gnomoniaceae, Clypeosphaeriaceae, Valsaceae, Melanconidaceae, and Melogrammataceae of the Sphaeriales. Traverso (**1906, 1907, 1913**) arranged his taxa differently: the Valsaceae included genera with stroma or pseudostroma while in the Ceratostomataceae and the Sphaeriaceae were diaporthaceous genera which had nonstromatic perithecia.

In the first of modern rearrangements, von Höhnel (**1917**) included in the family Diaporthaceae: subfamily I Eu-Diaportheen — genera with nonallantoid ascospores, subfamily II Valseen — genera with allantoid ascospores. The nonstromatic as well as stromatic taxa were placed in the first subfamily; ascospore septation and shape provided the major separation of genera. The second group included stromatic taxa only. In **1918**, in a series of "Fragmente," von Höhnel changed the name Valseen to the family Allantosphaeriaceae, in which he also included diatrypaceous genera, and amplified his 1917 concepts. Wehmeyer (**1926**) compared the arrangements of genera set forth by Lindau and by von Höhnel, and proposed a different realignment of allantoid-spored Ascomycetes — Allantosphaeriaceae for diatrypaceous forms and Diaporthaceae for diaporthoid forms. The Diaporthaceae included the stromatic taxa of von Höhnel's (**1917**) Eu-Diaportheae and Valseae. Only incidental mention was made of nonstromatic taxa.

Nannfeldt (**1932**), in his reorganization of ascomycete classification, recognized the orders Diaporthales and Valsales for von Höhnel's Eu-Diaportheen and Allantosphaeriaceen II Valseen, respectively. He removed the Diatrypaceae from close association with Valsales.

Luttrell (**1951**) described the Diaporthe type of centrum and the Endothia type of ascus to circumscribe the order Diaporthales. He included in the order two families, Diaporthaceae (including Valsaceae) and Melanosporaceae, and also considered the Ceratostomataceae in his discussion. Von Arx and Müller (**1954**) recognized the order Diaporthales with families Diaporthaceae and Valsaceae, as did Gäumann (Wynd translation, **1952**; **1964**). Munk (**1953**, **1957**) incorporated the Diaporthaceae, including Valsaceae, into the order Sphaeriales and recognized three subfamilies. Chadefaud (**1960**) recognized Diaporthales with Diaporthaceae or Valsaceae, Melanconidaceae, and Gnomoniaceae, based on characteristics of stromatic tissues. He regarded the Ceratostomaceae as closely related. Müller and von Arx (**1962**) considered the Diaporthaceae, including the Valsaceae, as a family in the Sphaeriales. They recognized two groups of genera in the didymosporous Diaporthaceae: one in which the ascus bases deliquesce and asci float free in the centrum, the other in which the ascus remains attached to the wall of the centrum. The classifications of Dennis (**1960**, **1968**), Kreisel (**1969**), Kobayashi (**1970**) and Müller and von Arx (**1973**) contained the Diaporthaceae in a broad sense in the Sphaeriales. In Wehmeyer's (**1975**) classification, the order Diaporthales included three families: the Gnomoniaceae for nonstromatic forms, and, for the stromatic forms, the Diaporthaceae with varied shapes of ascospores and the Valsaceae with allantoid ascospores.

THE ORDER DIAPORTHALES

The concept of the order Diaporthales presented in this contribution is based upon that of Luttrell (**1951**) for the family Diaporthaceae. Those Euascomycetes in which the perithecial centrum is pseudoparenchymatous not forming true paraphyses (Diaporthe type of Luttrell), are arranged in the subclass Parenchemycetidae of Barr (**1976a**) in five orders. The superficial, obligately parasitic members of the Erysiphales and Meliolales are separated readily from the Diaporthales. The three other orders, Diaporthales, Sordariales, and Coronophorales, are more closely related and have many features in common. In addition to the diagnostic features related below, habit is used to separate the Diaporthales from the Sordariales and Coronophorales. The fungi accepted here as belonging to the Diaporthales are parasitic or saprobic, never hypersaprobic, nor coprophilous.

The centrum of members of the Diaporthales is pseudoparenchymatous in early stages of development; paraphysis-like broad bands of cells may remain in the centrum, but typically no such structures are present in mature perithecia. The asci are elliptic or broadly oblong, occasionally cylindric, and when mature usually deliquesce at the base to fill the centrum and eventually are extruded through the ostiolar pore, although the asci remain attached to the subhymenium in a number of genera when mature. The apparently nonfunctional, refractive apical annulus of the ascus stains blue in nigrosin, a chitinoid reaction, and is nonamyloid. The external aspect of the perithecial wall is quite consistent throughout the order — the cells of the peridium form a textura epidermoidea (FIG. 2) in surface view. This is readily observable in the species with simple perithecia, but may be obscured in other species by surrounding stromatic tissues. Typically an elongate beak is formed and the ostiolar canal is lined by periphyses; the surface of the beak is composed of vertically oriented, elongate cells forming textura porrecta (FIG. 2). The main variable features of taxonomic value are utilized in the classification as diagrammed in FIG. 1: absence or presence and type of stromatic tissues, position of perithecia and beak in relation to the substrate, and shape and septation of ascospores.

Conidial states are known for a number of species in several genera of the Diaporthales. For the most part, the conidial state is enteroblastic-phialidic, with the conidiogenous cells produced in acervuli or in pycnidium-like locules in stromatic tissues. In a number of genera, the conidial state is holoblastic and the conidiogenous cells are annellated.

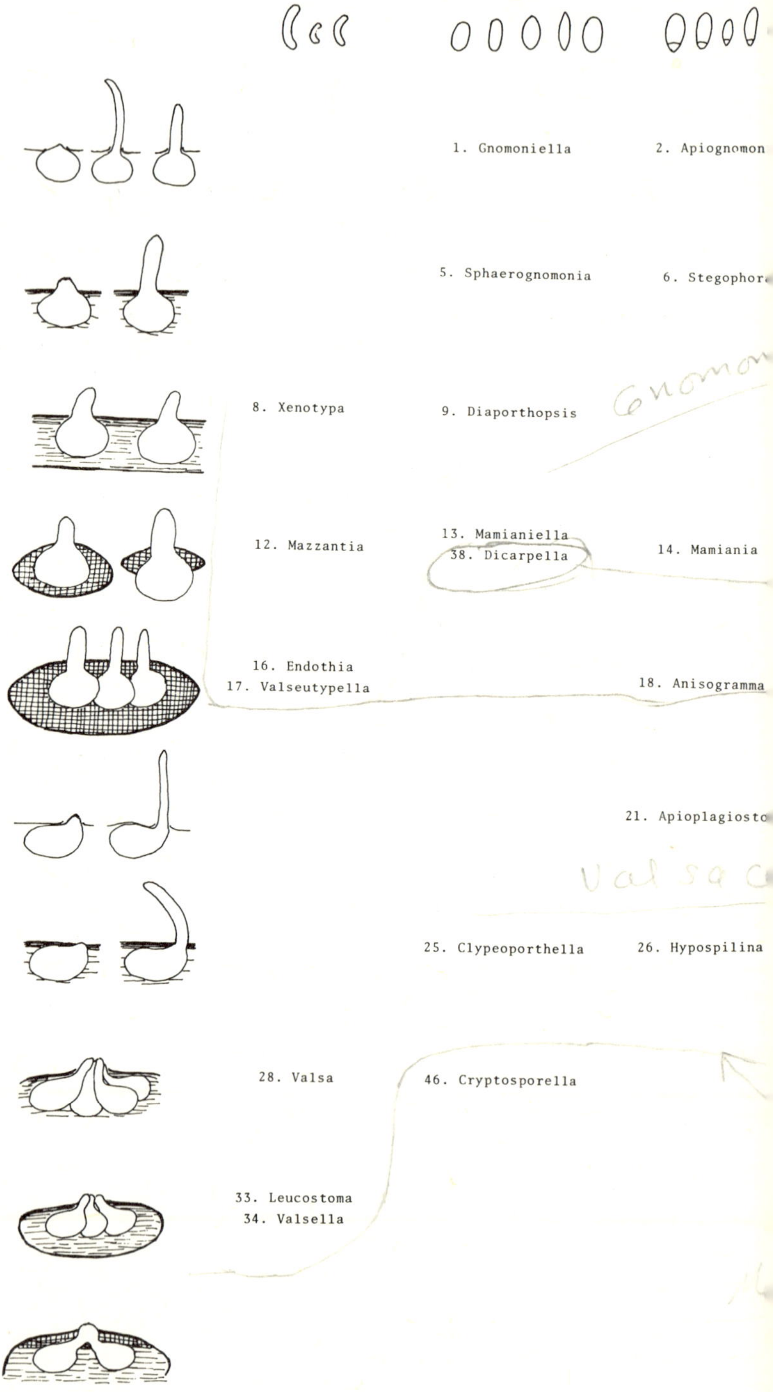

FIG. 1. Classification of genera of the Diaporthales according to position of perithec
and stromatic tissues, and ascospore shape and septation. Diagrammatic representatior

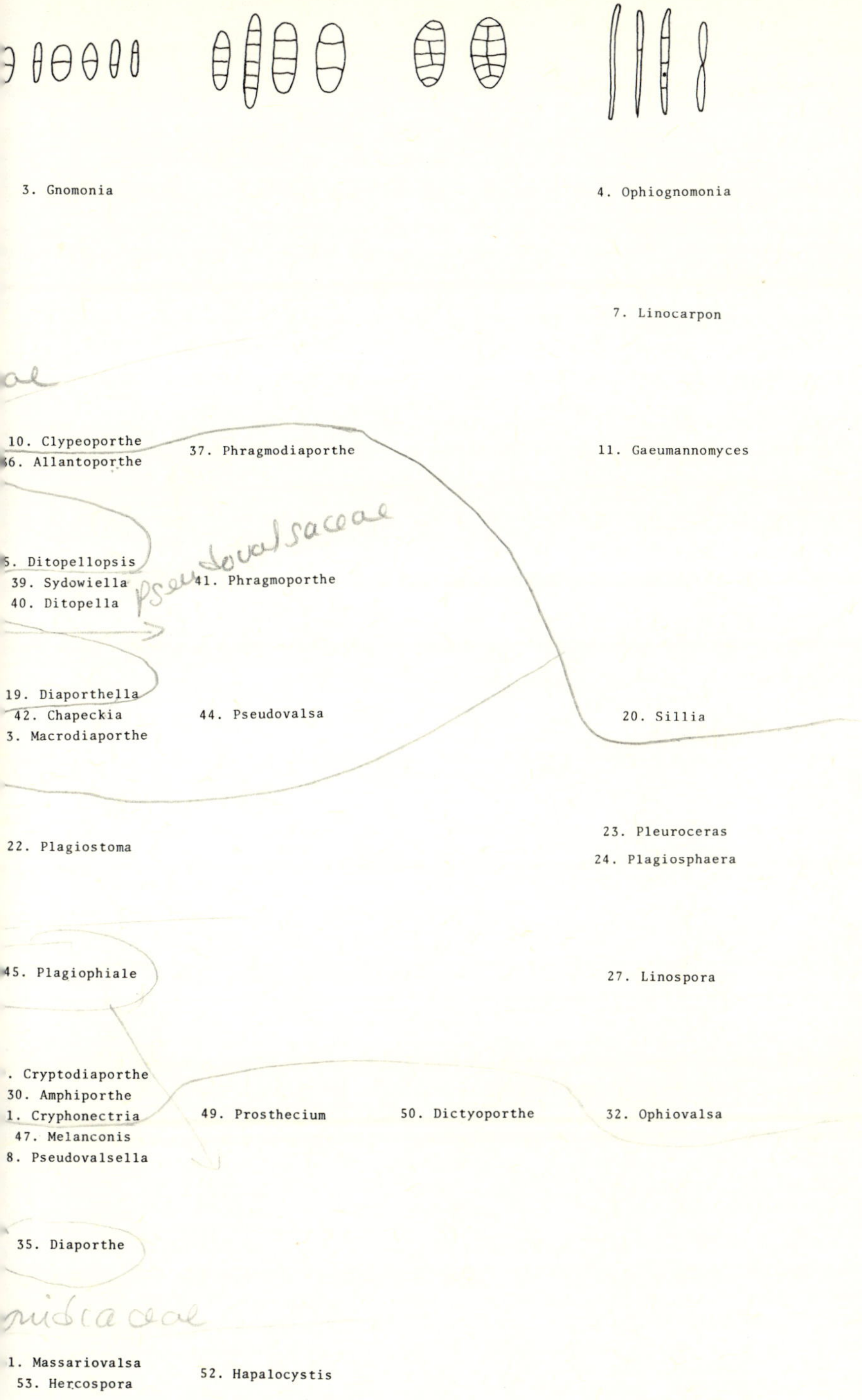

ıot to scale. In left hand column, horizontal lines indicate prosenchymatous stromatic
issues, cross-hatched areas indicate pseudoparenchymatous stromatic tissues.

The outline of classification presented in the following pages includes those genera of the Diaporthales, of which representatives are recognized to occur in temperate North America. In a few cases, genera not yet known from North America but which are expected to occur, are included. Under the detailed circumscription of tribes are inserted brief notes on extralimital genera for the use by investigators in other regions. Detailed treatments of the species of *Gnomonia* and of a number of closely related segregate genera are presented. Other genera are given less thorough treatment. Monographic studies do exist for a few genera, e.g. *Diaporthe*; for others sufficient information is not yet available.

For details on distribution, I am indebted to literature source as well as to many herbarium collections. The specimens cited are arranged for territories, provinces, and states, from north to south and east to west. The herbaria in which specimens are deposited are cited in accordance with Index of Herbaria (Holmgren and Keuken, **1974**).

To conserve space, numbered collections are cited only by collector or herbarium and number. If any possibility of confusion exists, details of locality, date, and collector are included.

CLASSIFICATION OF THE DIAPORTHALES

Diaporthales Nannfeldt, Nova Acta Reg. Soc. Sci. Upsal. ser. IV. 8(2): 53. 1932.

Valsales Nannfeldt, Nova Acta Reg. Soc. Sci. Upsal. ser. IV, 8(2): 54. 1932.

Perithecia small to large, immersed, erumpent, rarely superficial, with or without stromatic tissues; mycelium scanty or abundant, hyphae hyaline or brown, septate, branched, at times forming interwoven prosenchymatous stromatic tissues or compact-celled pseudoparenchymatous stromatic tissues; beaks usually elongate, short at times, occasionally lacking, ostiolar canal periphysate; centrum pseudoparenchymatous; cells of peridium a textura epidermoidea; asci unitunicate, thin walled, usually loosening from subhymenium and floating free in centrum, with refractive, chitinoid, nonamyloid apical annulus, aparaphysate or with paraphysoid tissue in the form of broad elongate cellular strands, soon deliquescent; ascospores variable in shape, septation and pigmentation, bipolarly symmetric usually, radially symmetric or asymmetric; conidial states variable; parasitic or saprobic.

Key to Suborders and Families of the Diaporthales

1. Perithecia small to medium sized; peridium thin, beaks narrow or broad; asci numerous, bases deliquescing early and asci freed and filling centrum; ascospores narrow or broad, hyaline, yellowish at times, rarely brownish in age, thin walled . . . . Suborder Gnomoniineae 2

1. Perithecia medium to large sized; peridium thin or thick; beaks broad; asci relatively few, bases usually remaining attached to subhymenial layer or at times deliquescing and asci freed and filling centrum; ascospores usually broad, hyaline, yellowish brown or dark brown, firm and usually thick walled . . . . Suborder Melanconidineae 3

2. Perithecia upright; beaks central, rarely eccentric, erumpent separately . . . . Family Gnomoniaceae, p. 18

2. Perithecia oblique or horizontal; beaks oblique or lateral, erumpent separately or converging through stromatic disc . . . . Family Valsaceae, p. 100

3. Perithecia upright; beaks central, rarely eccentric, erumpent separately . . . . Family Pseudovalsaceae, p. 151

3. Perithecia oblique or horizontal; beaks oblique or lateral, erumpent separately or converging through stromatic disc or fused together . Family Melanconidaceae, p. 176

Outline of classification of the Diaporthales

A. Suborder Gnomoniineae

I. Family Gnomoniaceae

I-1. Subfamily Gnomonioideae

I-1a. Tribe Gnomonieae

Genera: 1. *Gnomoniella* 2. *Apiognomonia* 3. *Gnomonia* 4. *Ophiognomonia*

I-2. Subfamily Stegophoroideae

I-2a. Tribe Stegophoreae

Genera: 5. *Sphaerognomonia* 6. *Stegophora* 7. *Linocarpon*

I-2b. Tribe Diaporthopsieae

Genera: 8. *Xenotypa* 9. *Diaporthopsis* 10. *Clypeoporthe* 11. *Gaeumannomyces*

I-3. Subfamily Mamianioideae

I-3a. Tribe Mamianieae

Genera: 12. *Mazzantia* 13. *Mamianiella* 14. *Mamiania* 15. *Ditopellopsis*

I-3b. Tribe Endothieae

Genera: 16. *Endothia* 17. *Valseutypella* 18. *Anisogramma* 19. *Diaporthella* 20. *Sillia*

II. Family Valsaceae

II-1. Subfamily Plagiostomoideae

II-1a. Tribe Plagiostomeae

Genera: 21. *Apioplagiostoma* 22. *Plagiostoma* 23. *Pleuroceras* 24. *Plagiosphaera*

II-2. Subfamily Valsoideae

II-2a. Tribe Linosporeae

Genera: 25. *Clypeoporthella* 26. *Hypospilina* 27. *Linospora*

II-2b. Tribe Valseae

Genera: 28. *Valsa* 29. *Cryptodiaporthe* 30. *Amphiporthe* 31. *Cryphonectria* 32. *Ophiovalsa*

II-2c. Tribe Diaportheae

Genera: 33. *Leucostoma* 34. *Valsella* 35. *Diaporthe*

B. Suborder Melanconidineae

III. Family Pseudovalsaceae

III-1. Subfamily Allantoporthoideae

III-1a. Tribe Allantoportheae

Genera: 36. *Allantoporthe* 37. *Phragmodiaporthe*

III-2. Subfamily Pseudovalsoideae

III-2a. Tribe Ditopelleae

Genera: 38. *Dicarpella* 39. *Sydowiella* 40. *Ditopella* 41. *Phragmoporthe*

III-2b. Tribe Pseudovalseae

Genera: 42. *Chapeckia* 43. *Macrodiaporthe* 44. *Pseudovalsa*

IV. Family Melanconidaceae

IV-1. Subfamily Melanconidoideae

IV-1a. Tribe Plagiophialeae

Genus: 45. *Plagiophiale*

IV-1b. Tribe Melanconideae

Genera: 46. *Cryptosporella* 47. *Melanconis* 48. *Pseudovalsella* 49. *Prosthecium* 50. *Dictyoporthe*

IV-2. Subfamily Massariovalsoideae

IV-2a. Tribe Massariovalseae

Genera: 51. *Massariovalsa* 52. *Hapalocystis*

IV-2a. Tribe Hercosporeae

Genus: 53. *Hercospora*

THE GENERA OF DIAPORTHALES

A. Suborder **Gnomoniineae** Barr, subordo nov.

Perithecia amplitudine parva vel media, raro magna, peridiis tenuibus, rostris angustis vel latis; asci ellipsoidi vel oblongi, raro cylindroidi, fundis deliquescentibus mature, ergo asci liberi et copiose centro; ascosporae hyalinae vel interdum flavae, raro demum brunneae, pariete tenuis. Genus typicus: *Gnomonia* Ces. & de Not.

I. Family **Gnomoniaceae** Winter in Rabenh., Krypt.-Fl. **1**(2): 570. 1886, Barr, emend.

Perithecia upright, beaks central, rarely eccentric, erumpent separately; ascospores nonseptate, one septate, occasionally several septate.

Key to Subfamilies and Tribes of Gnomoniaceae

1. Perithecia single in substrate, stromatic tissues not formed .... Subfamily Gnomonioideae, p. 18
One tribe, Gnomonieae

1. Perithecia single or grouped in substrate, stromatic tissues formed 2

2. Stromatic tissues prosenchymatous, forming a clypeus or interwoven hyphae or blackened zone or stromatic disc at surface of substrate, ill defined within substrate .... Subfamily Stegophoroideae . . 3

2. Stromatic tissues pseudoparenchymatous around one or several perithecia.. Subfamily Mamianioideae . . 4

3. Perithecia usually single in substrate, clypeus formed ...... Tribe Stegophoreae, p. 65

3. Perithecia separate or grouped, surrounded by interwoven hyphae or covered by blackened zone or stromatic disc..................... Tribe Diaporthopsieae, p. 73

4. Perithecia single or few in small stroma................ Tribe Mamianieae, p. 82

4. Perithecia few to many in well developed stroma .... Tribe Endothieae, p. 93

I-1. Subfamily **Gnomonioideae**

Perithecia single in substrate, stromatic tissues not formed; in leaves and petioles of deciduous trees and shrubs, herbaceous stalks, leaves, petioles, peduncles, rarely ferns or lycopods. (one tribe: Gnomonieae).

I-1a. Tribe **Gnomonieae**

Key to Genera of Gnomonieae

1. Ascospores nonseptate, ellipsoid, ovoid, or fusoid.................. 1. *Gnomoniella*
1. Ascospores septate.. 2
   2. Ascospores elongate, narrowly clavoid, septate in supramedian region .... 4. *Ophiognomonia*
   2. Ascospores variable in shape but not elongate clavoid......................... 3
3. Ascospores septate near base, apiosporous...................... 2. *Apiognomonia*
3. Ascospores septate in median, or supramedian or submedian region, not apiosporous.. .... 3. *Gnomonia*

1. **Gnomoniella** Sacc., Michelia **2**: 312. 1881; Syll. Fung. **1**: 413. 1882.

Stromatic tissues not formed, hyphae sparse in substrate; perithecia immersed in host tissues, upright, globose depressed, medium sized, beaks central, broad, short to elongate, straight or bent near apex, not formed at times; peridium narrow, consisting of several layers of large, compressed cells, darkened externally. Asci ellipsoid, numerous, apical annulus conspicuous, eight or four spored. Ascospores hyaline, ellipsoid, ovoid, or fusoid, usually inequilateral, ends rounded or pointed, one celled, cytoplasm containing one or two globules or minutely guttulate, wall firm, smooth, at times elongated at basal end into short broad appendage, overlapping biseriate or uniseriate in the ascus.

Conidial state where known: *Cylindrosporella* v. Höhnel (= *Asteroma* DC. ex Fr., Sutton, **1977**): Acervuli subcuticular; conidiophores phialidic; conidia hyaline, cylindroid or ellipsoid, one celled.

In overwintered leaves, petioles, peduncles of woody plants.

Type species: *Gnomoniella tubaeformis* (Fr.) Sacc.

The genus *Gnomoniella* contains few species and these are readily separable on the bases of ascospore shape and size. The medium-sized perithecia with stout beaks often cause deformation of the affected plant part.

Many names were transferred to *Gnomoniella* from *Gnomonia* by Saccardo (**1882**) because the ascospore septum was not observed, or was not mentioned by early investigators. These obligate nomenclatural synonyms are listed under the applicable species. *Gnomoniella curvicol-*

la (Peck) Sacc., Syll. Fung. **1**: 417. 1882, basionym: *Sphaeria curvicolla* Peck, Annual Rep. New York State Mus. **31**: 50. 1879; type—on *Polygonum articulatum* L. (*Polygonella articulata* (L.) Meisn.), Center, New York, Oct. C. H. Peck (NYS), is a species of *Eutypa*. **Eutypa curvicolla** (Peck) Barr, comb. nov.

Key to Species of *Gnomoniella*

1. Ascospores broadly ellipsoid, 11-15.5 × 6-9 μm. . . . . 1. *G. tubaeformis*
1. Ascospores ellipsoid or fusoid, 3.5-5.5(-7.5) μm wide 2
    2. Ascospores ellipsoid, 9-12 μm long, appendage basal, short and blunt. 2. *G. vagans*
    2. Ascospores fusoid, 12-18.5 μm long, without appendage . . . . 3. *G. hyparctica*

1. **Gnomoniella tubaeformis** (Fr.) Sacc., Syll. Fung. **1**: 413. 1882 FIG. 2

Sphaeria tubaeformis Fries, Syst. Mycol. **2**: 516. 1823.

Ceratostoma tubaeformis (Fr.) Ces. & de Not., Comment. Soc. Critt. Ital. **1**: 226. 1863.

Gnomonia tubaeformis (Fr.) Auersw. in Gonnerm. & Rabenh., Mycol. Eur. **5/6**: 22. 1869; Fuckel, Symb. Mycol. p. 120. 1870.

Perithecia 300-500μm diam, 160-350 μm high, thickly scattered, beak usually erumpent hypophyllously, rarely not formed, 200-450 μm long, 100-120 μm wide near base, slightly tapered to rounded apex. Asci 42-65 × (10.5-)13.5-18 μm. Ascospores 11-15.5 × 6-9 μm, hyaline, broadly ellipsoid, tapered to ± pointed ends, inequilateral.

Conidial state: *Cylindrosporella alnea* (Lev.) v. Höhnel (Klebahn, **1908**).

In overwintered leaves and petioles of *Alnus* spp., Europe, North America.

Material examined: *Europe:* Mus. Bot. Stockholm Herb. Mykol., Romell 15609 (GA); Mus. Bot. Stockholm Plantae Scand., Eliasson (GA). *North America: Alaska:* Haines, 29 Aug 1923, J.P. Anderson (BPI), Berg, 17 Aug 1948, C.L. Lefebre (BPI); *Maine:* Barr 1643 (MASS); *Massachusetts:* Barr 5598 (MASS); *Colorado:* Buzzard Creek, 24 Jun 1930, R.W. Davidson, Grand Mesa Mt., 12 Jul 1930, R.W. Davidson (both BPI); *Georgia:* GA 6703, 6705, 6706 (GA); *California:* Calif. Fungi 459 (BPI, GA, MICH, NY, TENN).

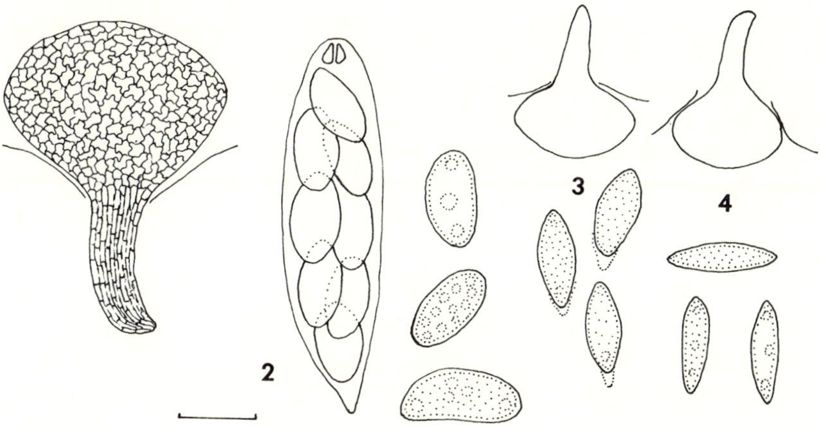

FIGS. 2-4. Species of *Gnomoniella*. 2. *G. tubaeformis,* outline of perithecium, ascus, ascospores. 3. *G. vagans,* outline of perithecium, ascospores. 4. *G.. hyparctica,* outline of perithecium, ascospores. Scale =100 μm (for perithecium) or 10 μm (for ascus and ascospores).

Specimens from California (Calif. Fungi 459) have light reddish brown beaks which are surrounded by loosely interwoven hyphae over the parallel upright hyphae of the beak peridium.

Gnomoniella tubaeformis begins development in living leaves and caused leaf spotting. Klebahn (**1908**, **1918**) attempted to culture this species but was unable to obtain growth, only germination of ascospores on culture media. The ascospores, when sprayed on leaves, caused spotting on which the conidial state developed. Conidia did not germinate.

Weese (**1935**) and Petrak (**1941**) corrected the citation of the species authority which was given by Saccardo and other authors as "Tode ex Fries."

Gnomoniella alnobetulae Volkart in Rübel, Engl. Bot. Jahrb. **47**: 517. 1912, is a species closely related to *G. tubaeformis.* The isotype specimen (Vestergren's Micromycetes rar. sel. 1144, on *Alnus viridis* DC., Graubünden, Fürstenalp, Switzerland, 11 Jun 1905, BPI) shows a fungus of similar sizes and aspect of perithecia and asci. However, the asci of *G. alnobetulae* are four spored and the larger ascospores measure 14.5-18.5 × 7.5-11(-13) μm. Another collection of *G. alnobetulae* from Switzerland is from Kt. Graubünden, Bivio, 12 Jun 1955, coll. E. Müller (TRTC).

2. **Gnomoniella vagans** Johans., Öfver. Kongl. Vet. Akad. Förh. **1884**(9): 163. 1884. FIG. 3

Perithecia 245-315 μm diam, 190-225 μm high, beak 150-210 μm long, 75-82 μm wide near base. Asci 50-63 × 9-13.5 μm, (4-7)8-spored. Ascospores 9-12 × (3.5-)4.5-5.5 μm, ovoid or ellipsoid, usually straight, occasionally inequilateral, wall prolonged beyond basal end into a short appendage ca. 2 μm long and 1 μm wide.

In peduncles and petioles of *Dryas octopetala* L., Iceland, Europe, *D. integrifolia* Vahl., North America.

Material examined: *Europe:* Lapponia Lulensis, Sweden, Aug 1900, T. Vestergren (GA, TRTC). *North America: Northwest Territories:* head of Clyde Inlet, Baffin Island, 16 Jul 1950, P. Dansereau (DAOM, MASS).

This species has the aspect and growth habit of *G. hyparctica*, and like it appears to be rare. *Gnomoniella vagans* was described from and reported twice from Iceland, and otherwise is known from the above collections.

3. **Gnomoniella hyparctica** (Lind) Barr, Contrib. Inst. Bot. Univ. Montreal **73**: 76. 1959. FIG. 4

Gnomonia hyparctica Lind, Meddel. om Grønland **71**: 176. 1926.

Perithecia 200-385 μm diam, 128-200 μm high, beak up to 300 μm long, 70-185 μm wide near base. Asci 57-81 × 10.5-16 μm. Ascospores 12-18.5 × 3.5-5.5(-7.5) μm, obovoid fusoid, inequilateral, tapered to both ends.

On overwintered peduncles and sepals of *Cassiope tetragona* (L.) D. Don., Europe, North America.

Material examined: *North America: Northwest Territories:* Coral Harbour, Southampton Island, Dist. of Keewatin, 16 Aug 1959, D.B.O Savile et al. (DAOM 70439); *Newfoundland:* Hebron, Labrador, 10 Jul 1954, R.T. Wilce (DAOM, MASS).

Lind originally described the fungus from Greenland and later (**1928**) reported two collections from Svalbard (Spitzbergen). Holm (**1975a**) did not find *G. hyparctica* on *Cassiope* in Scandinavia, and concluded that it was a truly arctic fungus. From the phanerogamic collections in UPS, he added to the few records known two more from Spitzbergen and one more from Canada (Northwest Territories, District of Mackenzie).

2. **Apiognomonia** v. Höhnel, Ber. Deutsche Bot. Ges. **35**: 632. 1917, nom. nud.; Ann. Mycol. **16**: 51. 1918.

Stromatic tissues not formed, hyphae sparse in substrate; perithecia immersed, upright, globose depressed, small to medium sized, beaks short papillate or conic to elongate, narrow or broad, central, rarely eccentric; peridium narrow. Asci ellipsoid, numerous, apical annulus shallow or conspicuous, eight spored. Ascospores hyaline, narrowly ellipsoid or fusoid, usually inequilateral, apiosporous, lower cell 1/4 to 1/5 size of upper cell, appendages short and narrow when formed, overlapping biseriate in the ascus.

Conidial state: parasitic, usually forming leaf spots (anthracnoses), producing acervuli or pycnidia; conidiophores phialidic; conidia hyaline, ellipsoid or fusoid, one celled. Microconidia also formed. Described as *Gloeosporium* Desm. & Mont. or *Discula* Sacc. spp.

In leaves of dicotyledonous trees and shrubs.

Type species: *Apiognomonia veneta* (Sacc.) v. Höhnel.

The main distinguishing character of *Apiognomonia* is the decided inequality of the two cells of the ascospores. The upper cell is three to five times longer than the lower cell. A number of species in *Gnomonia* have unequally septate ascospores, but the ratio of upper to lower cell is more in the range of 2:1 than 3:1, 4:1, or 5:1, and thus those ascospores are not truly apiosporous.

Several species of *Apiognomonia* are recognized in this study, although von Arx (**1951**) united a number of these under the earliest epithet, *A. errabunda*, and in **1957** grouped various conidial states on different host plants as *Discula quercina* (West.) von Arx. These species are indeed quite similar, as Klebahn (**1918**) had observed when he suggested that *Gnomonia platani*, *G. quercina*, and *G. tiliae* were perhaps "species sorores." However, Neely and Himelick (**1967**), in reference to the species on *Platanus* and *Quercus*, pointed out morphological and cultural differences, as well as noting failure of cross infection. Certainly host specificity seems to be the best way to recognize the species; in the key below the characters of beak width (but not length), perithecial size, and ascospore size and shape are utilized. The conidial states of the species on *Platanus* and *Quercus* may also be distinguished morphologically (Neely and Himelick, **1965**, **1967**). Neely (**1968**) described development of a somatic pressure cushion during formation of thin-walled pycnidia in living leaves.

Species not yet known from North America include *A. erythrostoma* (Pers. ex Fr.) v. Höhnel, which develops in leaves of species of *Prunus* in

Europe. The beaks are reddish brown and the ascospores 17-20 × 6 μm. Brooks (**1910**) and Likhité (**1926**) described perithecial development in this species. *Apiognomonia austriaca* Petrak was described from stalks of *Euphorbia* in Europe. The species has large ascospores, 23-27(-30) × 5-6.5 μm, and the lower cell is only 5-6 μm long.

Key to Species of *Apiognomonia*

1. Beaks relatively broad, perithecia medium sized (mostly 200-450 μm diam) 2

1. Beaks relatively narrow, perithecia small (mostly less than 200 μm diam); ascospores relatively broad (length: width ± 3:1) 6

 2. Ascospores relatively narrow (length:width ± 4:1) 3

 2. Ascospores relatively broad (length:width ± 3:1) 4

3. Ascospores 13-15(-20) × 3-4(-6) μm . . . . 4. *A. quercina*

3. Ascospores 15-17 × 3.5-4 μm . . . . 5. *A. errabunda*

 4. Ascospores (12-)16-20 × (4-)5-7 μm . . . . 1. *A. veneta*

 4. Ascospores smaller, (8-)10-13.5(-16) × (2.5-)3.5-5.5(-6) μm 5

5. Ascospores with basal end pointed, apical end pointed often 2. *A. alniella* var. *alniella*

5. Ascospores with both ends rounded . . . . 3. *A. alniella* var. *ribis*

 6. Ascospores 12.5-17.5 × 3.5-5.5 μm . . . . 6. *A. tiliae* var. *tiliae*

 6. Ascospores 11-15 × 3-4.5 μm . . . . 7. *A. tiliae* var. *magnoliae*

1. **Apiognomonia veneta** (Sacc. & Speg.) v. Höhnel, Ann. Mycol. **16**: 51. 1918. FIG. 5

Laestadia veneta Sacc. & Speg., Michelia **1**: 351. 1881.

Apiospora veneta Sacc. (in litt.) in Klebahn, Zeitschr. f. Pflanzenkr. **12**: 257. 1902.

Gnomonia veneta (Sacc. & Speg.) Klebahn, Jahrb. Wiss. Bot. **41**: 533. 1905. non *G. veneta* Speg. 1879.

Guignardia veneta (Sacc. & Speg.) Trav., Flora ital. Crypt. **2**: 392. 1906.

Apiosporiopsis veneta (Sacc. & Speg.) Mariani, Atti Soc. Ital. Sci. Nat. **50**: 165. 1911.

Gnomonia platani Klebahn, Deutsch. bot. Gesellsch. **1**: 28. 1914.

Perithecia 130-300(-430) μm diam, immersed in leaf tissues, beaks

50-100 μm long, 50-60 μm wide near base. Asci 45-64 × 13-21 μm. Ascospores (12-)16-20 × (4-)5-7 μm.

Conidial state: *Discula platani* (Peck) Sacc. sub *D. quercina* (West.) v. Arx (**1957**); *Gloeosporium platani* Oud. (Neely and Himelick, **1965**).

Causing leaf, petiole, and twig anthracnose of *Platanus* spp., Europe, North America, widespread. Numerous collections of the conidial state are deposited in various herbaria.

Material examined: (ascus state) *Europe:* Myc. venet. 1266 (MICH).

Klebahn (**1905**) provided a detailed description of the fungus. He germinated ascospores and obtained the conidial state in culture. Neely and Himelick (**1965, 1967**) gave additional information.

2. **Apiogonomoniella alniella** (Karst.) v. Höhnel, Hedwigia **62**: 49. 1921. var. **alniella** FIG. 6

Gnomonia alniella Karst. Mycol. Fenn. **2**: 119. 1873.

Gnomoniella albo-maculans Neger, Naturw. Z. f. Forst- und Landw. **10**: 432. 1912.

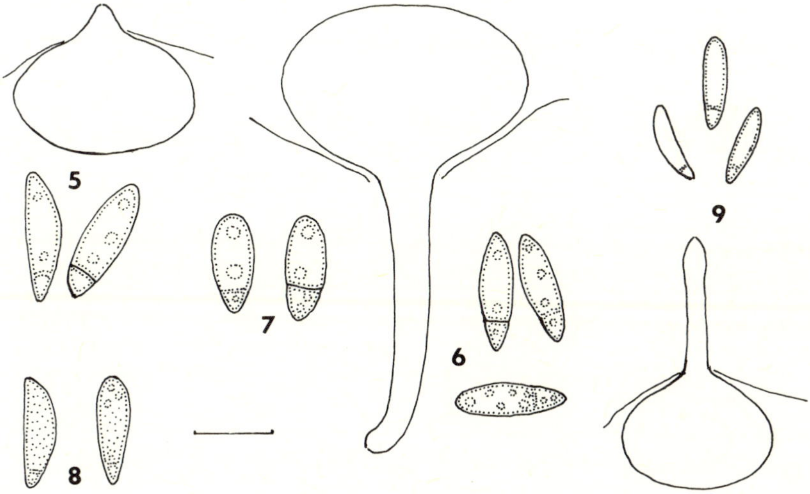

FIGS. 5-9. Species of *Apiognomonia*. 5. *A. veneta,* outline of perithecium, ascospores. 6. *A. alniella* var. *alniella,* outline of perithecium, ascospores. 7. *A. alniella* var. *ribis,* ascospores. 8. *A. quercina,* ascospores. 9. *A. tiliae* var. *magnoliae,* outline of perithecium, ascospores. Scale = 100 μm or 10 μm.

Perithecia (250-)315-450 μm diam, (150-)210-350 μm high, depressed, immersed, mostly hypophyllous, also in petioles, beaks central, curved, 200-500 μm long, 65-100 μm wide near base, tapered to 45-50 μm wide near tip. Asci (27-)40-54 × 7-12 μm. Ascospores 10-16 × 3.5-5.5 μm, ellipsoid fusoid with basal or both ends pointed, lower cell 2-5 μm long, 2.5-3 μm wide, at times with short blunt appendages.

Conidial state not formed.

On overwintered leaves of *Alnus* spp., Europe, North America.

Material examined: *Europe: Finland:* Karelia australis, Sippola, Ruotila, 7 Jun 1947, V. Kujala (DAOM 105437), Sülinjarvi, Poljä, Hoikin pwro, 29 Aug 1944, L. Korkonen (DAOM 105438); *Sweden:* Vesterbotten, Umeå, Jun 1909, J. Vleugel (TRTC). *North America: Quebec:* Manitou Gorge, 12 Jun 1955, R.T. Wilce (MASS, DAOM); *Maine:* Barr 3322b, 3540c (MASS).

Klebahn (**1918**) attempted infection experiments with ascospores on living leaves of *Alnus incana* (L.) Moench. and *A. glutinosa* (L.) Gaertn. The latter species was not infected. On *A. incana* brownish spots were formed, and on these white *Asteroma*-like mycelial mats developed. No conidia were produced. After two periods of wetting and drying the spotted leaves in September and November, perithecia developed by the end of November, and contained mature asci and ascospores by 20 December. Attempts to culture the species were unsuccessful because ascospores did not germinate on artificial media.

3. **Apiognomonia alniella** var. **ribis** Barr, var. nov. FIG. 7

Similis var. *alniella* vel ascosporae 10.5-13.5 × 4-6 μm, ellipsoideae et extrema rotundata.

Holotypus in foliis anni praeteriti *Ribis hirtellum* var. *saxosum*, "Mont Albert, Gaspe Provincial Park, 7 July 1957," a. M.E. Barr n. 1925b lecto (MASS).

Perithecia 275-460 μm diam, beaks 300-540 μm long, 66-82 μm wide near base, tapering to 43-50 μm near tip, hypophyllous, curved. Asci 27-33 × 9 μm. Ascospores 10.5-13.5 × 4-6 μm, ellipsoid, tapered to base, both ends rounded, lower cell 3-4 μm long, 3-3.5 μ wide, no appendages seen.

In overwintered leaves of *Ribes hirtellum* Michx. var. *saxosum* (Hook.) Fern., North America.

Known only from the type collection.

4. **Apiognomonia quercina** (Kleb.) v. Höhnel, Hedwigia **62**: 48. 1821.
FIG. 8

Gnomonia quercina Klebahn, Haupt- und Nebenfr. Ascom. p. 178. 1918.

Perithecia 130-440 μm diam, beaks 130-440 μm long, 35-60 μm wide near base. Asci 38-52 × 8-13 μm. Ascospores 13-15(-20) × 3-4(-6) μm.

Conidial state: *Discula quercina* (West.) v. Arx p.p. (**1957**); *Gloeosporium quercinum* West. (Neely and Himelick, **1965**).

Causing leaf, petiole and twig anthracnose of *Quercus* spp., Europe, North America, widespread. Numerous collections of the conidial state are deposited in herbaria.

Material examined: *Ontario:* Nashville, York Co., 2 Jun 1953, R.F. Cain (TRTC); *Michigan:* Barr 742 (MASS); *Maryland*: Barr 4878 (MASS).

Klebahn (**1918**) obtained conidia in cultures of ascospores. Neely and Himelick (**1967**) gave details of growth and compared *A. quercina* with *A. veneta* (as *Gnomonia platani*).

5. **Apiognomonia errabunda** (Rob. in Desm.) v. Höhnel, Ann. Mycol. **16**: 51. 1918.

Sphaeria errabunda Rob. in Desm., Ann. Sci. Nat. ser. III, **10**: 355. 1848.

Sphaerella errabunda (Rob. in Desm.) Auersw. in Rabenh., Fungi Eur. n. 944; Bot. Zeitung **23**: 307. 1866.

Gnomonia errabunda (Rob. in Desm.) Auersw. in Rabenh., Mycol. Eur. v/**6**: 25. 1869.

Perithecia ca. 200-300 μm diam, beaks short, 60-100 μm long and wide, hypophyllous. Asci 40-45 × 10-12 μm. Ascospores 15-17 × 3.5-4 μm.

Conidial state: *Gloeosporium fagi* (Desm. & Rob.) West. sub *Discula quercina* in von Arx (**1957**).

In leaves and petioles of *Fagus* spp., Europe, North America.

Material examined: *New Hamphire:* Barr 3893a (MASS); *Massachusetts:* Barr 5595 (MASS).

6. **Apiognomonia tiliae** (Rehm) v. Höhnel, Hedwigia **62**: 48. 1921. var. **tiliae**

Laestadia errabunda (Rob.) Rehm f. *tiliae* Rehm, Hedwigia **41**: (202). 1902; Ascom. exs. no. 1426.

Gnomonia tiliae Klebahn, Haupt- und Nebenfr. Ascom. p. 202-205. 1918.

Perithecia 140-180(-275) μm diam; beaks 40-120 μm long, 20-30 μm wide near base, narrow. Asci 35-51 × 8-12 μm. Ascospores 12.5-17.5 × 3.5-5.5 μm.

Conidial state: *Gloeosporium tiliae* Oud. sub *Discula quercina* (West.) von Arx (**1957**).

On leaves and petioles of *Tilia* spp., Europe; probably also occuring in North America.

Material examined: *Europe:* Rehm Ascom. exs. 1426 (type).

Klebahn (**1918**) described this species in detail. He obtained the conidial state from cultures of ascospores.

7. **Apiognomonia tiliae** var. **magnoliae** Barr, var. nov. FIG. 9

Similis var *tiliae,* sed ascosporae minores, 11-15 × 3-4.5 μm.

Holotypus in foliis anni praeteriti *Magnoliae soulangeanae* Soul., "Campus, University of Massachusetts, Amherst, 22 May 1961" a M.E. Barr n. 2900 lecto (MASS).

Similar to var. *tiliae* but ascospores smaller, 11-15 × 3-4.5 μm.

Known only from the type collection.

3. **Gnomonia** Ces. & de Not., Comment. Soc. Critt. Ital **1**(4): 231-232. 1863.

Rehmiella Winter, Hedwigia **22**: 2. 1883.

Cryptoderis β Gnomoniopsis Sacc., Syll. Fung. **2**: 229. 1883.

Gnomonia III *Gnomoniopsis* Winter, Kryptogamenfl. **1**(2): 588. 1887.

Gnomoniopsis (Sacc.) Berlese, Icones Fung. **1**: 93. 1894, non Stoneman, 1898.

Melanopelta Kirschstein, Ann. Mycol. **37**: 113. 1939.

Rostrocoronophora Munk, Dansk Bot. Arkiv. **15**(2): 98. 1953.

Stromatic tissues not formed, hyphae sparse in substrate; perithecia immersed, upright, globose or depressed, small to medium sized, often collapsing cupulate from the base on drying, occasionally collapsing above around beak, beaks central or eccentric, usually elongate, narrow (-50 μm diam near base) or broad (60-120 μm diam near base), in some species erumpent through an accumulation of white or yellowish granular material which forms a collar around base of beak; peridium narrow, consisting of few layers of compressed cells, darkened externally, beak wall usually darkened below and pallid at tip. Asci numerous, ellipsoid, floating free, apical annulus shallow or conspicuous, (two, four) eight (many) spored. Ascospores hyaline, rarely brownish in age, ellipsoid or fusoid or cylindroid, narrow or broad, one septate, occasionally developing one or two additional septa in age, septum median or submedian or supramedian, contents usually with two globules per cell, wall smooth, appendages when formed narrow and pointed (setose) or short and blunt, overlapping biseriate or oblique or in one or two fascicles in the ascus.

Conidial state: typically acervulate; conidiophores phialidic; conidia hyaline, ellipsoid or fusoid, one celled or one septate. Described as *Zythia* Fr., *Discula*, *Marssoniella* v. Höhnel (= *Marssonina* Magn., Sutton, **1977**), or *Cylindrosporella* (= *Asteroma*, Sutton, **1977**).

Usually saprobic, a few species parasitic, in leaves and petioles of deciduous trees and shrubs, leaves, peduncles, or stalks of herbaceous plants.

Lectotype: *Gnomonia vulgaris* Ces. & de Not. =*G. gnomon* (Tode ex Fries) Schroeter.

Cesati and de Notaris (**1863**) listed several species in *Gnomonia*, from which the lectotype must be selected. *Gnomonia vulgaris* Ces. & de Not. (=*Sphaeria gnomon* Tode ex Fries), *G. setacea* (Pers. ex Fr.), *G. capreae* (DC.), *G. leptostyla* (Fr.), *G. adunca* (Roberg.), *G. ciliata* (Riess), *G. amoena* (Nees ex Fr.), and *G. schoenoprasi* (Rabenh.) comprised the species which they officially placed in *Gnomonia*, while several others were listed as probably belonging in the genus. Of these species, *G. capreae* is a species of *Linospora*, *G. adunca* is a questionable species, *G. ciliata* is *Pleuroceras cryptoderis* (Auersw.) v. Höhnel, and *G. schoenoprasi* is a species of *Mycosphaerella*. The other four species of *Gnomonia* according to Cesati and de Notaris are indeed species of this genus. *Gnomonia leptostyla* and *G. amoena* have never been suggested as lectotype species. Von Höhnel (**1917**) selected *G. vulgaris*, the firstnamed species, as lectotype. Müller and von Arx (**1962**) reduced *G. vulgaris* to synonymy with *G. setacea* and designated the latter species as lectotype; Kobayashi (**1970**) followed their disposition. Bolay (**1972**), on the other hand, retained *G. vulgaris* as lectotype separate from *G. setacea.* I believe that *G. vulgaris* and *G. setacea* are two recognizable taxa, readily separated by differences in ascospore shape and position of the septum, as well as in the aspect of perithecia. Either species could serve as lectotype of the genus, but since von Höhnel's choice was *G. vulgaris,* that choice is followed. Since *Gnomonia gnomon* is not a tautonym, the older epithet rather than *G. vulgaris* is retained for the lectotype.

Rehmiella Winter was based on the polysporous nature of asci in *R. alpina* (Wint.) Winter. According to Bolay (**1972**) the species is similar to *Gnomonia comari* except for this one feature. Bolay included *Rehmiella* in the synonymy of *Gnomonia.* Earlier von Höhnel (**1919**) included *Rehmiella* as a synonym of *Ditopella,* but *R. alpina* lacks the pseudoparenchymatous stromatic disc which surrounds the beak in *Ditopella.*

Gnomoniopsis (Sacc.) Berlese was raised to generic rank for those species whose ascospores finally develop additional septa, and was based upon *Sphaeria chamaemori* Fries. This condition appears to be of only occasional occurrence, and is not sufficient to separate a genus. *Gnomoniopsis* is regarded as a synonym of *Gnomonia* (von Arx, **1951**; Müller, **1957**; Müller and von Arx, **1962**). *Gnomoniopsis* Stoneman (Bot. Gaz. **26**: 101. 1898) is *Glomerella* Schrenk & Spaulding.

Melanopelta Kirschstein, with *M. saxonica* Kirschstein, was described as forming a slight clypeus in the substrate. Kirschstein's species is synonymous with *Gnomonia sanguisorbae* (Rehm) Müller, and Müller in Müller and von Arx, (**1962**) did not uphold Kirschstein's genus.

Rostrocoronophora Munk was described as a genus of the Coronophoraceae for *R. geranii* Munk, although Munk (**1953**) noted the similarity of *R. gerani* to species of *Gnomonia.* Müller and von Arx (**1962**) considered *R. geranii* as a synonym of *Gnomonia geranii* Hollos, and synonymized *Rostrocoronophora* under *Gnomonia.*

A few species described in *Gnomonia* belong in the Loculoascomycetes. *Sphaeria gallophila* Ell. & Ev. (in Herb.) (*Diaporthe gallophila* Ell. & Ev., Bull. Torrey Bot. Club **8**: 90. 1881) was excluded from *Diaporthe* by Wehmeyer (**1933**) and was placed as *Gnomonia rostellata* (Fr.) Nits. It is instead a species of *Didymella* hypersaprobic in old perithecia of (presumably) *Diaporthe* sp. *Gnomonia sabalicola* Earle, Bull. Torrey Bot. Club **25**: 361. 1898 was transferred to *Wettsteinia sabalicola* (Earle) Barr, Contrib. Univ. Mich. Herb. **9**: 548. 1972.

Key to Sections and Species of *Gnomonia*

1. Ascospores narrow (length:width ± 5-6:1) 2
1. Ascospores relatively broad (length:width ± 3-4:1) 23
   2. Ascospores ± fusoid, tapering to pointed ends 3
   2. Ascospores ± cylindroid or ellipsoid, tapering to rounded ends; septum median, appendages blunt when formed 13
3. Ascospore septum submedian, appendages not usually formed; beaks relatively wide (over 50 μm) . . . . . Section 3. *Angustispora*. 4
3. Ascospore septum median or supramedian, appendages setose when formed; beaks narrow (less than 50 μm) 6
   4. Beaks erumpent through and surrounded by white collar; ascospores 6.5-7.5 × 1.5-2 μm, not appendaged . . . . . 11. *G. waldsteiniae*
   4. Beaks not surrounded by collar 5
5. Ascospores not appendaged, 6.5-12(-13) × (1.5-)2-2.5(-3) μm . . . . . 9. *G. comari*
5. Ascospores appendaged, 7.5-10.5(-12) × 1.5-2.5 μm . . . . . 10. *G. artospora*
   6. Ascospore septum ± median . . . . . Section 2. *Seta* . 7
   6. Ascospore septum supramedian . . . . . Section 1. *Gnomonia* . 10
7. Beaks erumpent through and surrounded by white collar; ascospores 10-16 × 1.5-2 μm . . . . . 8. *G. petiolorum*
7. Beaks not surrounded by collar 8
   8. Ascospores (13-)16-25(-30) μm long . . . . . 7. *G. nerviseda*
   8. Ascospores smaller, (10-)11-16(-18.5) μm long 9
9. Ascospores relatively broad, (1.5-)2-3 μm wide . . . . . 6. *G. cerastis*
9. Ascospores narrow, 1-2 μm wide . . . . . 5. *G. setacea*
   10. Beaks erumpent through and surrounded by yellowish collar; ascospores 16-21 × 3-3.5 μm . . . . . 4. *G. rigniacensis*
   10. Beaks not surrounded by collar 11
11. Peridium bright vinaceous brown, visible as rosy area on leaf surface, beaks elongate, 385-1200 μm long; ascospores 15-25 × 1.5-2.5 μm . . . . . 3. *G. amoena*
11. Peridium brown, not visible as rosy area, perithecia widely erumpent and collapsing cupulate around beak 12
   12. Beaks straight or slightly curved; ascospores (10-)12-17.5(-22.5) × 1.5-2.5 μm . . . . . 1. *G. gnomon*
   12. Beaks curved, ± eccentric; ascospores 13-15(-17.5) × 1.5-3 μm . . . . 2. *G. ostryae*
13. Saprobic and maturing in overwintered tissues . . . . . Section 4. *Cylindrica* . 14
13. Parasitic and maturing in living tissues; asci usually less than eight spored . . . . . Section 5. *Macula* . 22
   14. Ascospores 10-16.5 μm long, narrowly ellipsoid 15
   14. Ascospores 15-30(-37) μm long, ± cylindroid 18
15. Perithecia small, up to 350 μm diam; ascospores 10-14 × 2-3(-3.5) μm, appendaged . . . . . 17. *G. similisetacea*

15. Perithecia medium sized, up to 500 μm diam; ascospores 11-16.5 × 2-3 μm 16

16. Ascospores not appendaged, biseriate in elongate asci........... 18. *G. triostei*

16. Ascospores appendaged, fasciculate in short asci 17

17. Ascospores 11-15 × 2-3 μm .... 19. *G. fasciculata*

17. Ascospores 13-16.5 × 2-3 μm................................ 20. *G. rhuicola*

18. Ascospores 22-25 × 2.5-3 μm, appendaged .... 14. *G. aesculi*

18. Ascospores wider, (2-)3-5.5 μm.................................... 19

19. Ascospores 15.5-21(-25) μm long, not appendaged........................... 20

19. Ascospores (16-)22-30(-37) μm long, appendaged 21

20. Beaks narrow, ascospores inequilateral or slightly curved....... 15. *G. linnaeae*

20. Beaks medium broad; ascospores straight.................. 16. *G. leptostyla*

21. Perithecia 180-440 μm diam, beaks up to 500 μm long, mostly hypophyllous......... .... 12. *G. caryae* var. *caryae*

21. Perithecia 500 μm diam, beaks up to 900 μm long, mostly epiphyllous .... 13. *G. caryae* var. *ribis*

22. Asci usually four spored; ascospores (15-)18-24(-27) × 3-4.5 μm, appendages short and blunt.. 21. *G. alni*

22. Asci usually two spored; ascospores 23-39(-51) × 4.5-9(-10.5) μm, not appendaged .... 22. *G. dispora*

23. Ascospores ± fusoid, tapering to pointed ends abruptly, appendages short, ± triangular; beaks broad.................................. Section 6. *Fusispora*

One North American species.................................. 23. *G. dryadis*

23. Ascospores ellipsoid, tapering to rounded ends, appendages setose when formed .. 24

24. Ascospore septum submedian; beaks narrow at base and enlarged toward apex or relatively broad .... Section 8. *Clava* . 25

24. Ascospore septum median; beaks narrow to relatively broad .... Section 7. *Latispora* . 30

25. Beaks erumpent through and surrounded by yellow collar; ascospores 6.5-7.5 × 2-2.5 μm, appendaged................................... 35. *G. dalibardae*

25. Beaks not surrounded by collar; ascospores larger.......................... 26

26. Asci four spored; ascospores 11-16 × 2.5-3.5 μm, appendaged..... 34. *G. misella*

26. Asci eight spored; ascospores smaller.................................. 27

27. Beaks short and wide, ± papillate; ascospores 7-10 × 1.5-2.5 μm....... 33. *G. rauii*

27. Beaks more elongate, usually flared below tip 28

28. Ascospores 7.5-10.5 × 3-4 μm, not appendaged .... 32. *G. clavulata*

28. Ascospores usually longer, appendaged 29

29. Ascospores (7.5-)9-11(-15) × (1.5-)2.5-3.5 μm................. 31. *G. gei-montani*

29. Ascospores (7.5-)9-12 × 2.5-4(-5) μm .... 30. *G. myricae*

30. Beaks erumpent through and surrounded by yellowish collar; ascospores 9-12 × 2.5-4 μm, appendaged .... 29. *G. pulcherrima*

30. Beaks not surrounded by collar.. 31

31. Asci four spored; ascospores 10-15.5 × 2.5-3.5 µm, appendaged; beaks broad .... 28. *G. rubi*

31. Asci eight spored 32

32. Beaks broad; ascospores 9-16.5 × 3-4.5 µm, appendaged ..... 27. *G. sibbaldiae*

32. Beaks narrow 33

33. Perithecia (250-)350-540 µm diam; ascospores (6.5-)8.5-12(-13) × 2-3.5(-4) µm, appendaged .... 26. *G. ribicola*

33. Perithecia smaller, 150-300(-330) µm diam; ascospores narrower, 9-12 × (1.5-)2-3 µm. 34

34. Ascospores not appendaged .... 24. *G. intermedia* var. *intermedia*

34. Ascospores appendaged .... 25. *G. intermedia* var. *alni*

Gnomonia section 1. **Gnomonia**

Ascospores narrow, fusoid, tapering to ± pointed ends, septum supramedian, appendages setose, in a fascicle in the ascus. Beaks narrow.

Type species: *Gnomonia vulgaris* Ces. & de Not. = *G. gnomon* (Tode ex Fries) Schroeter.

A number of European species, not yet known from North America, belong in this section, including the following: *Gnomonia pruni* Fuckel (Symb. Mycol. p. 122. 1870) on leaves of *Prunus*, has perithecia 240-450 µm diam and ascospores 11-15 × 1-1.5 µm. *Gnomonia rosae* Fuckel (Symb. Mycol. p. 122. 1870) on leaves of *Rosa*, has perithecia 240-330 µm diam and ascospores 13-16(-21) × 1-2 µm, slightly larger than those of *G. pruni.*

1. **Gnomonia gnomon** (Tode ex Fries) Schroeter in Cohn, Kryptogamenfl. Schlesien **3**(2): 390. 1894. FIG. 10

Sphaeria gnomon Tode ex Fries, Syst. Mycol. **2**: 517. 1823.

Cryptospheria gnomon (Tode ex Fries) Grev., Flora Edinb. p. 360. 1824.

Gnomoniella gnomon (Tode ex Fries) Magnus, Pilze von Tirol, p. 490. 1905; House, New York State Mus. Bull. **205-206**: 35. 1918.

Gnomonia vulgaris Ces. & de Not., Comment. Soc. Critt. Ital **1**(4): 232. 1863.

Gnomoniella vulgaris (Ces. & de Not.) Sacc., Syll. Fung. **1**: 416. 1882.

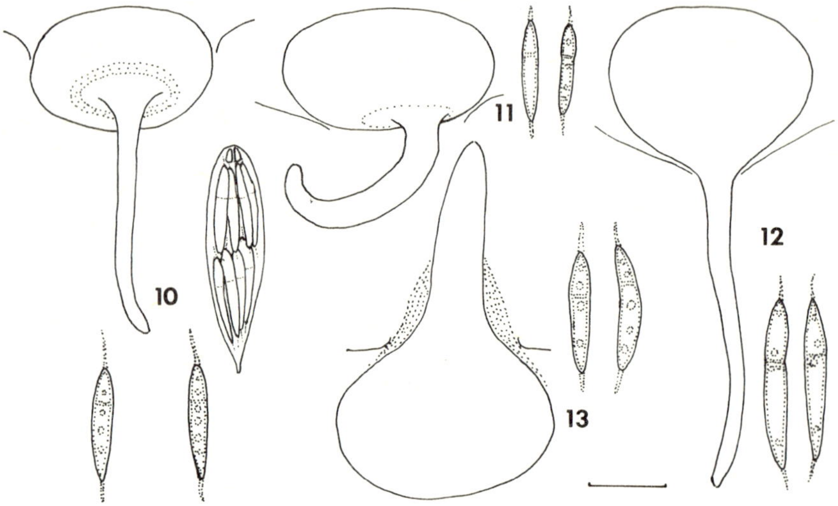

FIGS. 10-13. Species of *Gnomonia* section *Gnomonia.* 10. *G. gnomon,* outline of perithecium, ascus, ascospores. 11. *G. ostryae,* outline of perithecium, ascospores. 12. *G. amoena,* outline of perithecium, ascospores. 13. *G. rigniacensis,* outline of perithecium, ascospores. Scale =100 μm or 10 μm.

Perithecia immersed at first, becoming widely erumpent and appearing superficial, collapsing cupulate around base of beak on drying, scattered in indeterminate brownish blotches in living leaf, mostly hypophyllous in leaf blades or near veins, depressed globose, 180-275 μm diam; beaks central, narrow, straight or curved, (130-)180-410 μm long, 26-40 μm wide near base, tapering slightly to 25-35 μm near apex, at times slightly enlarged below apex; peridium dark brown, of about two layers of large cells, 15-20 μm wide. Asci numerous, 25-38 × 5-8 μm, ellipsoid fusoid, often with narrow, tapered stipe portion. Ascospores (10-)12-17.5 × 1.5-2.5 μm, hyaline, narrowly ellipsoid fusoid, straight or inequilateral, septum usually supramedian, occasionally nearly median, with two globules in each cell, appendages narrowly setose, 3-5 μm long, in one or two overlapping fascicles in the ascus.

On overwintered leaves of *Corylus* spp., Europe, North America.

Material examined: *Europe:* Mougeot & Nestler, Stirpes Crypt. Vogeso-Rhenanae 379 (DAOM); Sydow, Mycol. germ. 482 (TRTC); Heallsburg, Sweden, 19 Jun 1892, A.G. Eliasson (DAOM, GA). *North America: Ontario:* TRTC 10693, 10692, 2589 (TRTC); *Maine:* Barr 5833b (MASS); *Wisconsin:* H.C. Greene 3025 (DAOM 97142); *Georgia:* GA 6689 (GA); *California:* Barr 5954a (MASS); H.E. Parks 6064 (GA).

Miller (**1941**) reported this species on leaves of *Alnus* and *Ostrya* from Georgia, but the specimens belong to *G. setacea* and to *G. veneta* instead.

Klebahn (**1918**) and Morgan-Jones (**1959**) cultured *C. gnomon* and obtained perithecia in culture. Although *Gloeosporium coryli* (Desm. & Rob.) Sacc. has been suggested as the conidial state, none was produced, either in culture or when living leaves were inoculated.

According to the original description of *Sphaeria ischnostyla* Desm. (Ann. Sci. Nat. Bot. ser. 3, **11**: 357. 1849), in *Carpinus* leaves, Desmazieres' species is probably identical with *G. gnomon.* Fuckel's interpretation as *Gnomonia ischnostyla* (Desm.) Fuckel (Symb. Mycol. p. 121. 1870) differs, however. His fungus on *Acer pseudoplatanus* is *G. cerastis.* Saccardo (**1882**) suggested that *G. ischnostyla* was a variety of *G. setacea*, but Desmazieres had already separated *Sphaeria ischnostyla* from *S. setacea* on the basis of smaller perithecia with narrower beaks, pointed asci and linear, narrow ascospores.

2. **Gnomonia ostryae** de Notaris, Sferiacei Italici, cent. **I**: 42. 1863.
FIG. 11

Gnomonia veneta Speg., Michelia **1**: 457. 1879, non *G. veneta* (Sacc. & Speg.) Klebahn, 1905.

Perithecia immersed, hypophyllous in leaf blades, collapsing cupulate, depressed globose, 220-400 μm diam, beaks usually eccentric, narrow, curved or straight, 220-495 μm long, 36-50 μm wide near base, scarcely tapered, enlarged below pallid apex; peridium dark brown, 15-30 μm wide. Asci 27.5-40 × 6-9 μm. Ascospores 13-15.5(-17.5) × 1.5-3 μm, narrowly ellipsoid fusoid, straight to slightly curved, septum supramedian, appendages setose, 3-4.5 μm long, in two overlapping fascicles in the ascus.

In overwintered leaves of *Ostrya carpinifolia* Scop., Europe, *Ostrya virginiana* (Mill.) K. Koch and once on *Carya tomentosa* Nutt., North America.

Material examined: *Vermont:* Barr 4190 (MASS); *Georgia*: GA 6690, 6691, 6692, 6693 (as *G. gnomon*, GA); GA 6616 (on *Carya*, with other species, GA).

Although the eccentric position of the beak suggests *Plagiostoma*, the perithecia are upright as are the asci within, and this species must remain in *Gnomonia.* The name of *G. ostryae* surely refers to the same fungus as *G. veneta.* They are given as separate species in Traverso's

(**1906**) compilation of Italian fungi. Descriptions are similar; the only difference used in the key is the presence of setose appendages on ascospores of *G. veneta* and the lack of appendages in *G. ostryae.* This character alone is not sufficient, especially since the presence of appendages on ascospores has been overlooked in many species.

3. **Gnomonia amoena** (Nees ex Fries) Ces. & de Not., Comment. Soc. Critt. Ital. **1**(4): 232. 1863. FIG. 12

Sphaeria amoena Nees ex Fries, Syst. Mycol. **2**: 519. 1823.

Gnomoniella amoena (Nees ex Fries) Sacc., Syll. Fung. **1**: 414. 1882.

Perithecia scattered in leaf blades or veins, immersed and raising leaf tissues, rosy vinaceous around base of beak, 330-575 μm diam, 275-375 μm high, beaks central, narrow, elongate, 385-1200 μm long, 32-60 μm wide near base, tapering then enlarged below pallid apex, vinaceous black; peridium bright vinaceous, pigment not leached in KOH. Asci 30-40 × 5-9 μm. Ascospores 15-25 × 1.5-2.5 μm, narrowly ellipsoid fusoid, straight to slightly curved, septum supramedian or nearly median, appendages setose, 3-5 μm long, in two overlapping fascicles in the ascus.

In overwintered leaves of *Corylus avellanea* L., Europe, *C. californica* (A. DC.) Rose, North America.

Material examined: *Europe:* Mougeot, Nestler & Schimper, Stirpes Cryptogamae 1251 (DAOM). *North America: Oregon:* near Sulphur Springs, Autumn 1914, G.B. Posey (TRTC, under the name and with perithecia of *Mamiania coryli*); Oregon Agric. Coll. 2646 (GA 6601, under the name and with perithecia of *Mamiania coryli*).

Fries' (**1823**) brief description of *Sphaeria amoena* stressed the elevated rosy disc perforated by the elongate beak, and occurrence on *Carpinus betula* and *Corylus avellanea.* Later authors (Auerswald, **1869**; Winter, **1887**) provided microscopic details. Fuckel (**1870**) gave a different interpretation of the species: he described the ascospores as biseriate, simple, 8 × 4 μm, and apparently had a fungus with white collar surrounding the beak (*Mamiania coryli*?) because he included *Sphaeria petiolorum* Schw. in synonymy.

4. **Gnomonia rigniacensis** Sacc. & Flag., Syll. Fung. **17**: 666. 1905.
FIG. 13

Perithecia immersed in veins of leaf, 300-360 μm diam, beak up to 380 μm long, around 80 μm wide near base, enlarged to 140 μm wide by collar of yellowish granular material. Asci 44-63 × 9-12 μm. Ascospores 16-21 × 3-3.5 μm, narrowly ellipsoid fusoid, straight or inequilateral, septum supramedium, appendages setose, 3-4 μm long, overlapping biseriate in the ascus.

In overwintered leaves of *Cornus rugosa* Lam., North America, *C. sanguinea*, Europe.

Material examined: *North America: Wisconsin:* Ferry Bluff, Prairie du Sac, Sauk Co., 7 May 1948, H.E. Greene 1261 (BPI).

The description of *G. rigniacensis* from France fits this collection except for the slightly shorter and broader ascospores which were said to have the septum submedian. However, Saccardo and Flageolet (in Saccardo, **1905**) compared the species to *G. cerastis*, *G. errabunda* and *G. petiolicola*. The ascospore septum is median in *G. cerastis*, near the base in *G. errabunda* (*Apiognomonia*) and supramedian in *G. petiolicola* (*Plagiostoma*).

Gnomonia section **Seta** Barr, sectio nov.

Ascosporae angustae, fusoideae, in extrema plus minusve acutae, gradatim descrescentiae, septum medium, appendiculae setaceae; rostra angusta. Species typica: *Gnomonia setacea* (Pers. ex Fries) Ces. & de Not.

Ascospores narrow, fusoid, tapering to ±pointed ends, septum ±median, appendages setose (bristle like), in one or two fascicles in the ascus. Beaks narrow.

Type species: *Gnomonia setacea* (Pers. ex Fries) Ces. & de Not.

5. **Gnomonia setacea** (Pers. ex Fries) Ces. & de Not., Comment. Soc. Critt. Ital. **1**(4): 232. 1863.
FIG. 14

Sphaeria setacea Pers. ex Fries, Syst. Mycol. **2**: 518. 1823.

Perithecia immersed, occasionally becoming erumpent, scattered, usually hypophyllous in leaf blade, or in veins or petioles, depressed globose, 150-300(-450) μm diam, 165-230 μm high; beaks central, straight or curved, narrow, 150-500 μm long, up to 1 mm or longer at times, 30-52 μm wide near base, slightly tapering to apex; peridium narrow, brown. Asci 24-33(-42) × 5-10 μ. Ascospores (10-)12-16(-18.5) × 1-2 μm, hyaline, narrowly fusoid, straight or inequilateral, septum median, with two globules in each cell, appendages usually narrow, setose, 2-9 μm long, occasionally blunt, 2-3.5 μm wide, in one or two overlapping fascicles in the ascus, or overlapping obliquely.

On overwintered leaves of various deciduous shrubs, trees, especially species of *Betula, Alnus, Quercus, Castanea,* occasionally on *Lycopodium* sp., cosmopolitan.

Material examined (representative collections only): *Europe:* Sydow, Mycotheca march. 2953 (NY under *Gnomonia subversa*). *North America: Newfoundland:* Hebron, Labrador, 10 Aug 1954, R.T. Wilce (MASS); *Quebec:* Barr 2042 (MASS); *Ontario:* Bear Island, Lake Temagami, 13 Jun 1931, R.F. Cain (TRTC, GA, DAOM); *Manitoba:* WINF(M) 3036b, 4582, 6787 (WINF(M)); *British Columbia:* Barr 467, 508 (UBC), 5583, 5582 (MASS); *Yukon Territory:* DAOM 71518 (DAOM); *Maine:* Barr 1660a (MASS); *New Hampshire:* Barr 3819b, 3888, (MASS); *Vermont:* Barr 2989 (MASS); *Massachusetts:* Barr 5285, 6062 (MASS); *New York:* Barr 1883 (MASS); *New Jersey:* Ell. & Ev. N.A.F. 2138 (MICH); *Delaware:* GA 6646 (GA); *North Carolina:* GA 6647, 6648 (GA); *Tennessee:* GA 6648, 6649, 6650 (GA); *Georgia:* GA 6611-6615, 6624-6643 (GA).

Klebahn (**1918**) cultured *Gnomonia setacea* f. *alni* Vleugel. He obtained conidia which were formed on short side branches of hyphae. Perithecia were also produced. Morgan-Jones (**1959**) obtained perithecia in culture, but no conidial state.

Several varieties and forms of *G. setacea* have been described, but some of these are separate species, others appear to be substrate variants only. This species is widespread in distribution and occurs on a larger number of plants than do most species of *Gnomonia.*

6. **Gnomonia cerastis** (Riess) Ces. & de Not., Comment. Soc. Critt. Ital 1(4): 233. 1863. FIG. 15

Sphaeria cerastis Riess in Rabenh., Herb. mycol. n. 1737; Hedwigia 1: 24. 1853 (sine descr., tab. III, fig. 2, m, n, o); Auersw. Mycol. Eur. Pyren. 5/6: 27. 1869.

Sphaeria petioli Fuckel, Fungi rhen. 537. 1866.

Gnomonia petioli (Fckl.) Cooke, Fungi brit. p. 103. 1866.

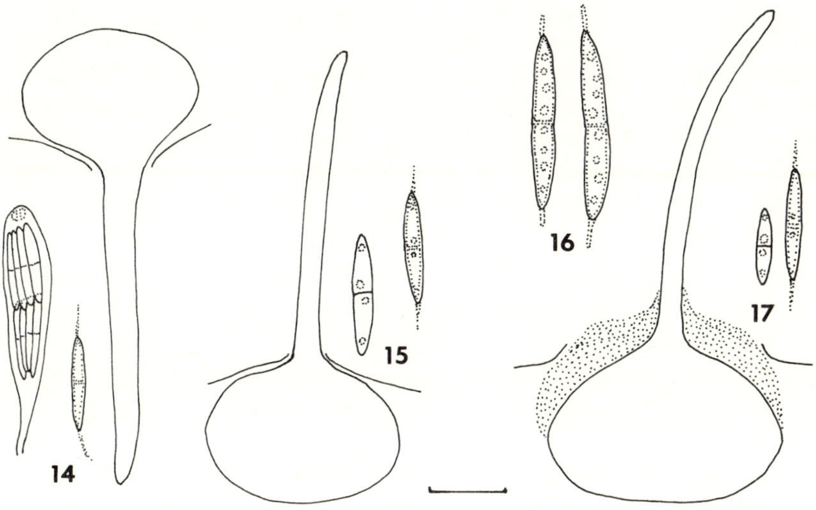

FIGS. 14-17. Species of *Gnomonia* section *Seta.* 14. *G. setacea,* outline of perithecium, ascus, ascospore. 15. *G. cerastis,* outline of perithecium, ascospores. 16. *G. nerviseda,* ascospores. 17. *G. petiolorum,* outline of perithecium, ascospores. Scale =100 μm or 10 μm.

Perithecia scattered in petioles, veins, and blades of leaves, depressed globose, (170-)240-450 μm diam, 130-210 μm high, beaks central to eccentric, narrow, short (40-50 μm) or elongate, 330-450(-1000) μm, 30-40 (-60) μm wide at base, tapering to apex. Asci 24-35 × 5-8.5(-10) μm. Ascospores 11-16×(1.5-)2-3 μm, narrowly fusoid ellipsoid, septum median, appendages setose when formed, overlapping biseriate or obliquely fasciculate in the ascus.

In overwintered leaves of *Acer* spp., Europe, North America.

Material examined: *Europe:* Herb. Barbey-Boissier 88; Ludwig, Flora von Westfalen (both DAOM, TRTC). *North America: British Columbia:* Barr 73, 501a, 509a (UBC); *Massachusetts:* Barr 4939 (MASS).

This species is closely related to *G. setacea* and differs in having slightly wider ascospores. It could perhaps be considered a form of *G. setacea*, but all of the specimens on *Acer* show slight size and shape differences of the ascospores from those of typical *G. setacea*.

7. **Gnomonia nerviseda** Cole, J. Agric. Res. **50**: 91-92. 1935. FIG. 16

Gnomonia setaceae var. *caryae* Dearn. & House, New York State Mus. Bull. **266**: 76. 1925.

Gnomonia caryae var. *pecanae* Cole, J. Agric. Res. **47**: 878. 1933.

Perithecia hypophyllous in blades, veins and petioles of leaves, depressed globose, 200-440 μm diam, scattered, beak 250-500(-800) μm long, 40-50 μm wide near base, tapering to 30 μm near apex, at times surrounded by narrow white collar at base. Asci 39-48 × 5.5-10 μm. Ascospores (13-)16-25(-30) × 2-2.5(-3.5) μm, hyaline, narrowly ellipsoid to cylindroid, or enlarged above and tapered downward, straight or inequilateral, septum median, with two globules in each cell, often bearing delicate setose appendages, 2-7.5 μm long, in two overlapping fascicles in the ascus.

Conidial state: *Leptothyrium nerviseda* Cole, *Leptothyrium caryae* var. *pecanae* Cole: Acervuli hypophyllous, depressed, 35-95 μm wide; conidiophores 14-18 × 2-4 μm; conidia 8-13 × 2-3 μm, hyaline, 1-celled.

On overwintered leaves of *Carya* spp., North America.

Material examined: *North America: New York:* Greenbush, Rensselaer Co., Jun, C.H. Peck (NYS, type of *G. setacea* var. *caryae*); *Maryland:* Shear 6008 (BPI); *Oklahoma:* Muskogee, Nov 1943, L.W. Larsh (BPI); *Georgia:* Albany, 22 Apr 1932, C.L. Shear (BPI); GA 6617, 6618, 6619, 6620, 6623, 6665 (GA); *Louisiana:* Barr 6337 (MASS), Ferriday, 30 Mar 1933, J.R. Cole (BPI, lectotype of *G. nerviseda*); Shreveport, 30 Apr 1933, J.R. Cole (BPI, lectotype of *G. caryae* var. *pecanae*); Shreveport, May 1930, J.R. Cole (BPI).

Cole termed the disease symptoms "vein spot" for *G. nerviseda* (**1935**) and "liver spot disease" for *G. caryae* var. *pecanae* (**1933**). The conidial state was produced in culture from both ascospores and conidia.

8. **Gnomonia petiolorum** (Schw. ex Fries) Cooke, Grevillea **7**: 54. 1878. FIG. 17

Sphaeria petiolorum Schw. ex Fries, Syst. Mycol. **2**: 518. 1823.

Gnomoniella amoena var. *petiolorum* (Schw. ex Fr.) Sacc., Syll. Fung. **1**: 414. 1882.

Perithecia immersed in petioles, major veins and in base of leaf blade, depressed globose, 275-330 μm diam, mostly hypophyllous; beaks emergent through and surrounded at base by collar of white crystalline material, single or up to three together, beaks narrow, straight or slightly curved, 600-825 μm long, 39-52 μm wide near base, tapering to apex, canal periphysate; peridium narrow, 10-15 μm wide, yellowish brown. Asci 22-27.5 × 5.5 μm, tapering to delicate stipe. Ascospores 10-16 × 1.5-2 μm, hyaline, ellipsoid fusoid, septum median, straight or slightly curved, often with delicate setose appendages 4-5 μm long, in one or two overlapping fascicles in the ascus.

On overwintered leaves of *Liquidambar styraciflua* L., occasionally on *Acer rubrum* L., North America.

Material examined: *Delaware:* Jackson 2159 (TRTC); GA 6677 (GA); *Maryland:* Shear 3928 (BPI); Suitland, Whetzel, Davidson et al. (BPI); *Tennessee:* GA 6678 (GA); *Georgia:* Ravenel, Fungi Amer. exs. 374 (BPI); GA 6668-6676 (GA); Luttrell, culture; Barr 6176 (MASS); *Alabama:* 2 Jun 1854, T.M. Peters (BPI); *Florida:* F 48576 (GA, MASS); *Louisiana:* Ell. & Ev. N.A.F. 2543 (MASS); Barr 6319 (MASS).

The species has been reported on *Fraxinus* from New York but the specimens so labelled in NYS are *G. artospora* instead. Perithecia are produced in culture (Luttrell, in litt.).

Gnomonia amoena, on *Corylus* leaves, is remarkable by the rosy disc over the perithecia and slightly broader ascospores with the septum supramedian. It does not possess the white collar of *G. petiolorum.*

Gnomonia section **Angustispora** Barr, sectio nov.

Ascosporae angustae, fusoideae, in extrema acuta plus minusve gradatim decrescentiae septum submedium, exappendiculae vel setaceae; rostra comparate lata. Species typica: *Gnomonia comari* Karst.

Ascospores narrow, fusoid, tapered to ±pointed ends, especially to basal end, septum submedian, usually without appendages, or appendages setose, overlapping biseriate or oblique in the ascus. Beaks relatively wide.

Type species: *Gnomonia comari* Karst.

9. **Gnomonia comari** Karsten, Mycol. Fennica 2(Pyrenomycetes): 122-123. 1873. FIG. 18

Gnomoniella comari (Karst.) Sacc., Syll. Fung. **1**: 415. 1882.

Gnomoniella guttulata Starb., Bihang till K. Svenska Vet.-Akad. Handl. **15**(III): 10. 1889.

Gnomonia guttulata (Starb.) Kirschst., Ann. Mycol. **33**: 219. 1935.

Apiognomonia guttulata (Starb.) Wehm., Canad. J. Res., C, **20**: 585. 1942.

Gnomonia agrimoniae Brefeld & v. Tavel, Untersuchungen aus dem Gesamtgebiet der Mykologie **10**: 232. 1891.

Gnomonia pusilla Sacc. & Flageolet, Syll. Fung. **17**: 664. 1905.

Gnomonia occulta Kirschst., Verh. Bot. Ver. Brandenburg **48**: 58. (1906) 1907.

Gnomonia herbicola A.L. Smith, Trans. Brit. Mycol. Soc. **3**: 221. 1910.

Gnomonia fragariae Klebahn f. sp. *fructicola* Arnaud, Traite de Pathol. veg. Encycl. Mycol. **4**(Paris): 1558-1562. 1931.

Gnomonia fructicola (Arnaud) Fall, Canad. J. Bot. **29**: 309. 1951.

Perithecia (180-)220-325(-400) μm diam, 180-250(-350) μm high, base often collapsing cupulate, beaks 100-325 μm long, 50-70 μm wide near base, tapered to apex. Asci 22-38(-52) × 4.5-7.5 μm. Ascospores 6.5-12(-13) × (1.5-)2-2.5(-3) μm, straight or inequilateral to slightly curved at times, septum submedian, appendages not formed, biseriate or oblique in the ascus.

Conidial state: *Zythia fragariae* Laibach (syn.: *Phyllosticta gei* Bres., *P. grandemaculans* Bub. & Krieger, *Gloeosporium fragariae* Arnaud).

In leaves, petioles, and stalks of several genera of Rosaceae, also in *Epilobium hirsutum* L., Europe, North America.

Material examined: *North America: Newfoundland: Potentilla* sp., 19 Jul 1957, R.T. Wilce (MASS); *Nova Scotia: Agrimonia* sp., Victoria Park, Truro, 7 Sep 1929, 23 Jun 1933, Wehmeyer (MICH); *Fragaria* sp., Kentville, 20 Jun 1952, D. Creelman (conidial) (DAOM 35014); *Quebec: Fragaria* sp., Stanstead, 7 Jul 1964, G.L. Brown (DAOM 45804); *Rubus chamaemorus* L., trail to McKenzie, 11 Jun 1955, R.T. Wilce (MASS); *New Jersey: Fragaria* sp., Elm, 17 May 1911, C.L. Shear (BPI).

Bolay (**1972**) has contributed significantly to knowledge of *G. comari*, especially to its physiological requirements in pure culture. He provided the synonymy for the species listed here and differentiated this from other taxa which occur on members of the Rosaceae. Earlier studies of

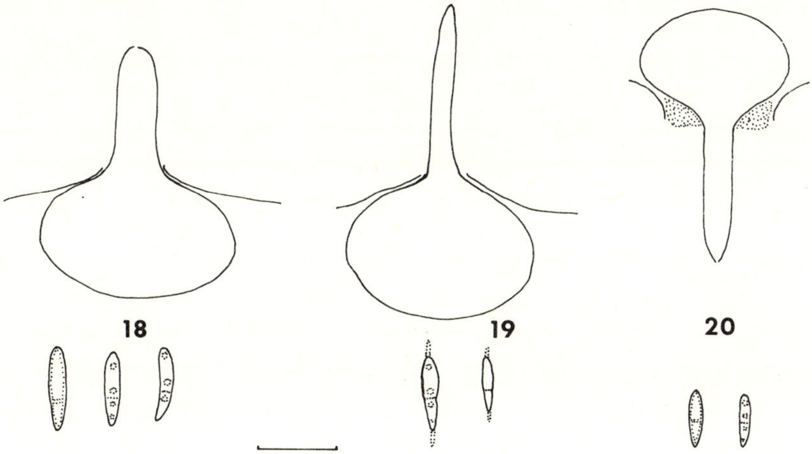

FIGS. 18-20. Species of *Gnomonia* section *Angustispora*. 18. *G. comari,* outline of perithecium, ascospores. 19. *G. artospora,* outline of perithecium, ascospores. 20. *G. waldsteiniae,* outline of perithecium, ascospores. Scale =100 μm or 10 μm.

the species, its conidial state, and the disease it causes on *Fragaria* in North America were made by Alexopoulos and Cation (**1948, 1952**) and by Fall (**1951**).

10. **Gnomonia artospora** (Dearn. & House) Dearn. & House, New York State Mus. Circ. **24**: 36. 1940. FIG. 19

Diaporthe artospora Dearness & House, Bull. New York State Mus. **205-206**: 48. 1919.

Gnomonia petiolophila f. *fraxini* Ell. & Ev. (in herb., ined.)

Perithecia 200-385 μm diam, 200-230 μm high, beaks 150-385 μm long, narrow, 39-50(-70) μm wide at base. Asci (15-)20-33 × (5-)6-9 μm. Ascospores 7.5-10.5(-12) × 1.5-2.5 μm, ±fusoid, often slightly curved, septum submedian, upper cell slightly wider than lower, appendages 1.5-2.5 μm long, overlapping biseriate to triseriate or oblique in the ascus.

On petioles of *Fraxinus nigra* Marsh., occasionally on *Castanea* sp., *Onoclea sensibilis* L., North America.

Material examined: *Ontario:* London, 3 Apr 1894, J. Dearness (type of *G. petiolophila* f. *fraxini*, NY, BPI); Gull Lake Portage, Lake Temagami, 25 Jun 1932, S.M. Pady (TRTC 3839, DAOM 85490); *New York:* Sandlake, May, C.H. Peck (NYS, type of *D. artospora*); Guilderland, May, C.H. Peck, Buffalo, 14 Feb, G.W. Clinton (as *G. petiolorum*, NYS); *Louisiana:* Lafayette, 21 Mar 1888, A.B. Langlois (on type of *Laestadia orientalis* Ell. & Ev., NY).

The collection on *Onoclea* was originally thought to be an undescribed species, but on comparison it could not be separated from *G. artospora.*

11. **Gnomonia waldsteiniae** Dearn. & House, New York State Mus. Circ. **24**: 37. 1940. FIG. 20

Perithecia 160-210 μm diam, 130-150 μm high, depressed globose, immersed, scattered to grouped in petioles and veins of leaf, beak erumpent hypophyllous, ±200 μm long, 50-60 μm wide at base, surrounded by a white collar; peridium 12-15 μm wide, two to three layers of light dull grayish or yellowish brown compressed cells. Asci 22.5-27 × 4.5-6 μm. Ascospores 6.5-7.5 × 1.5-2 μm, hyaline, ellipsoid fusoid, ends obtusely pointed, straight or slightly curved, septum submedian or nearly median, contents minutely guttulate, overlapping biseriate or oblique in the ascus.

On overwintered leaves and petioles of *Waldsteinia fragarioides* (Michx.) Tratt., North America.

Material examined: *New York:* Clarksville, 11 May 1924, H.D. House 143 (type, Dearness Herb. in DAOM).

Gnomonia section **Cylindrica** Barr, sectio nov.

Ascosporae angustae, ellipsoideae vel cylindroideae, in extrema rotundata gradatim decrescentiae, septum medium, exappendiculae vel appendiculae breves et obtusae; rostra angusta vel lata. Species typica: *Gnomonia caryae* Wolf.

Ascospores narrow, ellipsoid to ±cylindroid, slightly tapered to rounded ends, septum median, appendages short and blunt when formed, biseriate, triseriate, or obliquely fasciculate in the ascus. Beaks narrow or broad.

Type species: *Gnomonia caryae* Wolf.

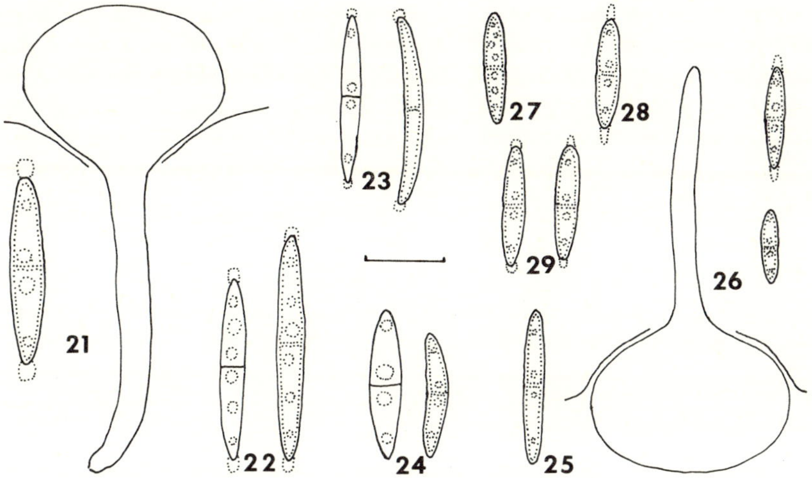

FIGS. 21-29. Species of *Gnomonia* section *Cylindrica*. 21. *G. caryae* var. *caryae*, outline of perithecium, ascospore. 22. *G. caryae* var. *ribis*, ascospores. 23. *G. aesculi*, ascospores. 24. *G. linnaeae*, ascospores. 25. *G. leptostyla*, ascospore. 26. *G. similisetacea*, outline of perithecium, ascospores. 27. *G. triostei*, ascospore. 28. *G. fasciculata*, ascospore. 29. *G. rhuicola*, ascospores. Scale =100 μm or 10 μm.

12. **Gnomonia caryae** Wolf, Ann. Mycol. **10**: 491. 1912. var. **caryae** FIG. 21

Gnomonia setacea var. *macrospora* Ell. & Ev., North Amer. Fungi 3429. 1896 (nom. nud.)

Perithecia 180-440 μm diam, 160-330 μm high, depressed globose, beaks usually erumpent through lower leaf surface, 200-500 μm long, 30-55(-80) μm wide near base, central or eccentric; peridium 15 μm wide. Asci 50-92 × (8-)10-15.5 μm, apical annulus 2-3.5 μm high. Ascospores (16-)22-30(-37) × (2-)3-5.5 μm, hyaline, ±cylindroid, septum median, appendages blunt, up to 3-5.5 μm long, up to 3-4.5 μm wide, in two fascicles in the ascus.

Conidial state: *Gloeosporium caryae* Ell. & Dearn., Proc. Acad. Sci. Phila. 82. 1891.

On overwintered leaves of *Carya* spp., North America.

Material examined: *North America: Massachusetts:* Barr 6209 (MASS); *New Jersey:* Ell. & Ev. N.A.F. 3429 (MASS, type of *G. setacea* var. *macrospora*); *Virginia:* Arlington Cemetery, Jan 1932, C.L. Shear (BPI); *Georgia:* GA 6616, 6621, 6622, one unnumbered specimen from Pulaski Co., 5 May 1940, D. Polkill (GA); *Alabama:* Auburn, May 1912, F.A. Wolf (BPI, type, NY). Various herbaria contain a number of collections of the conidial state.

Wolf (**1912**) described germination of ascospores and conidia and the production of acervuli on living leaves which were inoculated with ascospores.

13. **Gnomonia caryae** var. **ribis** Barr, var. nov. FIG. 22

Similis var. *caryae* sed peritheciis vel 500 μm diametro, rostro vel 900 μm longis.

Holotypus in foliis *Ribi cynosbati,* "McCarthy Field, Baxter State Park, Maine, 26 June 1962" a M.E. Barr n. 3230b lecto (MASS).

Perithecia 500 μm diam, 250 μm high, depressed globose, mostly epiphyllous, beaks up to 900 μm long, 60 μm wide at base, tapered to 40 μm wide at apex; peridium 24-30 μm wide. Asci 50-55 × 10-12 μm, apical annulus 3-3.5 μm high. Ascospores 26.5-30 × 3-4 μm, hyaline, cylindroid, septum median, appendages 3-4.5 μm long, 3 μm wide, in a fascicle in the ascus.

On overwintered leaves of *Ribes cynosbati* L., North America.

Known only from the type collection.

14. **Gnomonia aesculi** Oud., Beih. Bot. Centrabl. **11**: 527. 1902. FIG. 23

Perithecia 220-330 μm diam, 140-220 μm high, depressed globose, amphigenous, beaks 385-445 μm long, 26-40 μm wide at base and scarcely tapering to apex. Asci 40-48 × 7.5-10 μm. Ascospores 22-25 × 2.5-3 μm, narrowly ellipsoid cylindroid, septum median, appendages blunt, 1.5-2 μm long and wide, in oblique fascicle in the ascus.

In overwintered leaves of *Aesculus* spp., Europe, North America.

Material examined: *Georgia:* Savannah River, Elbert Co., 2 May 1938, J.H. Miller (GA 6711).

This collection has slightly longer ascospores than those described for *G. aesculi* (16-18.5 × 2.5 μm) but the narrow cylindroid shape and median septum reinforce the identification. The Georgia material bears delicate blunt appendages on each end of the ascospores, another feature not found (or not observed) in the original material. *Gnomoniella hippocastani* Brun., according to the description in Sylloge Fungorum **9**: 575, differs from *Gnomonia aesculi* in still smaller ascospores, 14-15 × 2-2.5 μm, and larger perithecia with thickened bases of beaks, which raise the petiole surface into pustules. *Gnomoniella hippocastani* from the description is suggestive of *Cryptodiaporthe petiolicola* (Peck) Barr.

15. **Gnomonia linnaeae** Auersw. in Rabenh. & Gonnerm., Mycol. Europ. **5/6**: 24. 1889. FIG. 24

Perithecia 250-500 μm diam, 370 μm high, globose or depressed, beak erumpent through upper leaf surface, 350-370 μm long or up to 1 mm, 50-60 μm wide at base, straight or bent; peridium brown. Asci 44-55 × 5.5-11 μm. Ascospores 15.5-20(-24) × 3.5-4(-5) μm, hyaline, cylindroid fusoid, ends obtuse or pointed, inequilateral to curved, septum median, overlapping biseriate in the ascus.

On overwintered leaves of *Linnaea borealis* L. var. *americana* (Forbes) Rehd., Europe, North America.

Material examined: *North America: Quebec:* Barr 1924 (MASS, overmature); *Maine:* Katahdin Lake, 22 Aug 1920, Yorks camp, 2 Aug 1940, both coll. C.L. Shear (BPI); *New Hampshire:* Randolph, Jul 2, 1932, S.K. Harris (MASS).

This fungus was described from Germany and has been reported from Spain, Sweden, and Lapland. It is said to be rare in Europe. The North American collections are sparse in numbers of perithecia. Petrak identified the two from Maine in BPI.

16. **Gnomonia leptostyla** (Fries) Ces. & de Not., Comment. Soc. Critt. Ital. **1**(4): 232. 1863. FIG. 25

Sphaeria leptostyla Fries, Syst. Mycol. **2**: 517. 1823.

Perithecia 200-350 μm diam, widely erumpent, hypophyllous, beaks 200-400 μm long, (45-)72-100 μm wide near base, tapering to 50-70 μm wide near apex, central or eccentric. Asci 45-70 × 8-15 μm. Ascospores 17-21(-25) × 2.5-4.5 μm, septum median, cylindroid, appendages not seen.

Conidial state: '*Marssonia*' *juglandis* (Lib.) Sacc.

On *Juglans* spp., Europe, North America.

Material examined: *Europe:* Herb. Barbey-Boissier 89 (TRTC). *North America*: *Ontario*: TRTC 31025, 31041 (TRTC); *Massachusetts:* Barr 2880a (MASS). Frequently collected as the conidial state.

Traverso (**1906**) utilized the pre-Friesian name *Sphaeria juglandis* DC., as *Gnomonia juglandis* (DC.) Trav. (Flora Ital. Crypt. **1**: 343. 1906). Klebahn (**1907**) described the conidial and perithecial states in detail. In culture, both conidia and ascospores produced both conidia and perithecia. Likhité (**1926**) described development of perithecia from infected leaf material.

17. **Gnomonia similisetacea** Barr, sp. nov. FIG. 26

Perithecia 165-350 μm diametro, 110-250 μm alta, globosa depressa, rostra 275-550 μm longa, 39-52 μm lata et 26 μm apicem versus. Asci (24-)40-45 × 6-10 μm. Ascosporae 10-14 × 2-3.5 μm, angustae ellipsoidae, septum medium, in extrema rotundata gradatim decrescentiae, appendiculae breves et obtusae.

In herbarum caulibus, vel ramorum vel foliis frutibus. Holotypus in foliis anni praeteriti *Euonymi alati,* "Conway, Massachusetts, 3 April 1974," a M.E. Barr n. 6199 lecto (MASS).

Perithecia 165-350 μm diam, 110-250 μm high, depressed globose, beaks 275-550 μm long, 39-52 μm wide near base and tapering to 26 μm near apex. Asci (24-)40-45 × 6-10 μm. Ascospores 10-14 × 2-3.5 μm, ellipsoid, septum median, upper cell slightly broader than lower, tapering to rounded ends, appendages 2.5-3 μm long, 2.5 μm wide, in two fascicles or triseriate in the ascus.

In overwintered leaves of *Euonymus alatus* (Thunb.) Sieb., *Vaccinium oxycoccos* L., and stalks of *Aralia nudicaulis* L., North America.

Material examined: *North America: Ontario:* on *Aralia,* Bear Island, Lake Temagami, 19 Jun 1932, R.F. Cain (TRTC); Lake of Two Rivers, Algonquin Park, 4 May 1968, R.F. Cain (TRTC); *Massachusetts*: on *Euonymus,* Barr 6199 (holotype, MASS); *Oregon:* on *Vaccinium,* Clatsop, 1925, H.F. Bain (BPI).

These collections, on quite different host plants, are remarkably similar in all respects. In perithecia and asci they could be referred to *G. setacea*, but ascospore shapes and appendages differ in the two species. In fact, Shear, Stevens and Bain (**1931**) did identify the Oregon specimen on *Vaccinium* as *G. setacea,* and earlier I had considered the identification correct. Now it is obvious that a taxon with the above characteristics has a relatively wide host range, just as does *G. setacea.*

18. **Gnomonia triostei** (Dearn. & House) Dearn. in Wehmeyer, The Genus *Diaporthe*, p. 270. 1933. FIG. 27

Diaporthe triostei Dearness & House, Bull. New York State Mus. **233-234**: 34. 1921.

Gnomonia triostei (Dearn. & House) Dearn. & House, Circ. New York State Mus. **24**: 37. 1940.

Perithecia 215-430 μm diam, globose, collapsing when dry, base seated on discolored plant tissues, beak short and barely protruding through epidermis or elongate, up to 430 μm long, 50-85 μm diam at base, tapering upward and enlarged below apex. Asci 39-50 × 6-8 μm. Ascospores 13.5-15 × 2-3 μm, narrowly ellipsoid, ends rounded, yellowish in mass but hyaline separately, straight to inequilateral, septum median, overlapping biseriate or oblique in the ascus.

In dead stems of *Triosteum aurantiacum* Bickn., North America.

Material examined: *North America: New York:* Glenmont, 16 May 1919, H.D. House (NYS, type).

19. **Gnomonia fasciculata** Fuckel, Symb. Mycol. p. 121. 1870. FIG. 28

Gnomoniella fasciculata (Fckl.) Sacc., Syll. Fung. **1**: 415. 1882.

Perithecia 200-540 μm diam, depressed, ca. 190 μm high, beaks narnow, 150-495 μm long, 40-52 μm wide near base and tapering to apex. Asci (22-)28-42 × (6.5-)8-10 μm. Ascospores (8-)11-15 × 2-3 μm, ellipsoid, septum median, ends rounded, appendages blunt, 2.5-3.5 μm long, in oblique fascicle or triseriate in the ascus.

In overwintered leaves of *Quercus* spp., Europe, North America.

Material examined: *North America: New Jersey:* Newfield, 15 Jun 1879, J.B. Ellis (NY, in specimen of *Venturia ciliata* var. *quercina* Ellis, ined.); *Georgia:* GA 6631, 6644 (with *G. setacea*) (GA); *Louisiana:* Barr 6333 (MASS); *Utah:* Barr 6141 (MASS); *Arizona:* Barr 6095 (MASS).

The ascospores of *Gnomonia fasciculata* were described as simple, with obtuse, not acuminate, ends, by Fuckel (**1870**). The median septum visible in the North American material was evidently overlooked. This species could be confused easily with *G. setacea*. The perithecia and asci tend to be somewhat larger than those of *G. setacea*, although size ranges overlap, but the ascospores furnish definite separation of the two species. In *G. fasciculata*, ascospores are narrowly ellipsoid with blunt appendages on the rounded ends, and are broader than the narrowly fusoid ones furnish with setose appendages in *G. setacea*.

20. **Gnomonia rhuicola** Barr, sp. nov. FIG. 29

Perithecia 180-500 μm diametro, rostra 350-650 μm longa, 50-60 μm lata et 40 μm apicem versus. Asci 35-45 × 7.5-10 μm. Ascosporae 13-16.5 × 2-3 μm, angustatae ellipsoidae, septum medium, in extremo rotundata gradatim decrescentia, appendiculae breves et obtusae.

In petiolis et folii *Rhui* spec. Holotypus: "Newfield, New Jersey, 3 and 4 July 1879," a J.B. Ellis lecto (NY).

Perithecia 180-500 μm diam, beaks 350-650 μm long, 50-60 μm wide near base and tapering to 40 μm wide near apex. Asci 35-45 × 7.5-10 μm. Ascospores 13-16.5 × 2-3 μm, narrowly ellipsoid, septum median, upper cell slightly wider than lower, tapering to rounded ends, appendages blunt, 2 μm long, 1 μm wide, ±fasciculate in the ascus.

In overwintered petioles and blades of leaves of *Rhus* sp., North America.

Known only from the type collection.

Ellis did not publish this species, but the specimen, composed of a mixture of petioles and leaflets, is sufficient to serve as type. The specific epithet is that given to the collection in Ellis's herbarium.

Gnomonia section **Macula** Barr, sectio nov.

Ascosporae ellipsoidae, angustae, in extrema rotundata gradatim decrescentiae, septum medium, exappendiculae vel appendiculae breves et obtusae; asci plerumque duo- vel quatuor-spori; rostra angusta; in foliis vivis. Species typica: *Gnomonia alni* Plowright.

Ascospores ellipsoid, relatively narrow, tapering to rounded ends, septum median, appendages short and broad when formed, oblique and less than eight in the ascus. Ostiole narrow. Parasitic in living leaves.

Type species: *Gnomonia alni* Plowright.

Gnomonia provancheriana Thümen in Provancher, Le Naturaliste Canadien **9**: 10. 1878 may belong in this section. The fungus was described from Quebec as forming straw-colored spots in living leaves of *Betula papyracea*. Asci were 35 × 10 μm and the ellipsoid ascospores, described as one-celled, measured 7 × 4 μm. This species was never recorded by Saccardo in any of the volumes of Sylloge Fungorum, and it is due to the courtesy of Dr. R. Pomerleau (in litt.) that I was made aware of it. No specimens have been seen.

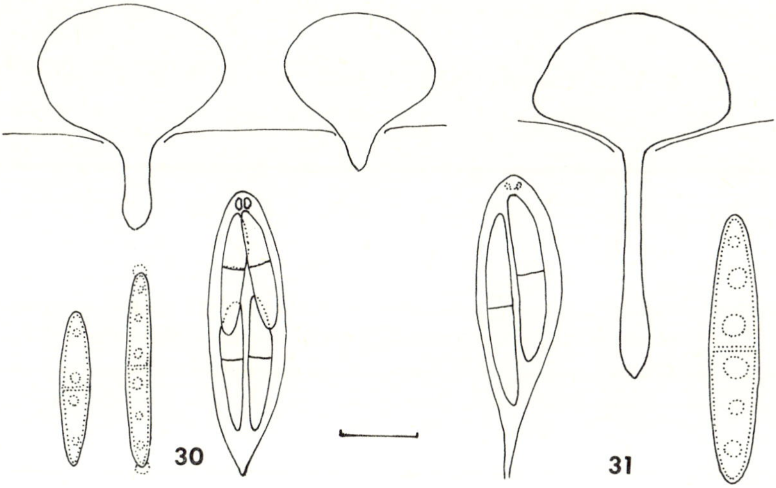

FIGS. 30, 31. Species of *Gnomonia* section *Macula*. 30. *G. alni*, outline of perithecia, ascus, ascospores. 31. *G. dispora*, outline of perithecium, ascus, ascospore. Scale =100 μm or 10 μm.

21. **Gnomonia alni** Plowright, Grevillea **7**: 74. 1878. FIG. 30

Hypospila californica Dearness & Barth. Mycologia **18**: 248. 1926.

Perithecia immersed in leaf tissues, numerous and thickly scattered in large irregular brown spots which originate at leaf margin and spread inward, bases visible on upper leaf surface, collapsing on drying, perithecia upright, globose depressed, 175-320 μm diam, 140-250 μm high, beaks erumpent hypophyllously, short papillate or somewhat elongate, 60-100(-200) μm long, 40-70 μm wide, tapering to apex or slightly flared below apex; peridium of light yellowish brown compressed layers of cells, darkened externally, 7.5-15 μm wide, yellowish brown hyphae in leaf tissues. Asci 33-45(-55) × 7.5-15 μm, ellipsoid, (2-, 3-)4-(5-, 6-) spored. Ascospores (15-)18-24(-27) × 3-4.5 μm, hyaline, light dull brownish in age, ellipsoid or narrowly obovoid, tapering slightly to rounded ends, straight or inequilateral, septum median, occasionally appendaged.

Conidial state: *Leptothyrium alni* Zeller: Acervuli subcuticular; conidiophores short, phialidic; conidia 6-7.5 × 1.5 μm, hyaline, narrowly ellipsoid, straight or curved, one celled, with one or two guttules.

Parasitic in living leaves of *Alnus* spp., western North America.

Material examined: (representative collections): *Washington:* Bellingham, 25 Aug 1916, J.R. Weir 3890 (BPI as *G. perversa*, NYS); *Oregon:* Fungi of Oregon 1032 (FLAS, NY, TRTC); Oregon Agric. College, 2 Jul 1915, G.B. Posey (TRTC); Corvallis, 29 Jul 1914, H.P. Barss (BPI, type of *Leptothyrium alni*); *California:* Riverside, 12 Aug 1924, E. Bartholomew (type and isotype of *Hypospila californica*, Dearness 5658, Dearness Herb. in DAOM); Ellis & Ev. N.A.F. 2360 (BPI, MASS, MICH, NY); Calif. Fungi 311 (NY, TRTC); Fungi Col. 231 (DAOM, NY); Baker, Plants of Pacific Slope 1848 (NY); *Arizona:* Wet Canyon, Mt. Graham, 4 Oct 1954, P.D. Keener (BPI); West Fork Oak Creek, Prescott, 21 Oct 1956, P.H. Keener (BPI).

22. **Gnomonia dispora** Demaree & Cole ex Barr, sp. nov. FIG. 31

Gnomonia dispora Demaree & Cole, Phytopathology **26**: 1028. 1936.

Perithecia 160-300 μm diametro, 120-200 μm alta, rostra 110-300 μm, angusta. Asci 40-55 × 9-13 μm, bispori. Ascosporae 23-39(-51) × 4.5-9(-10.4) μm, hyalinae, ellipsoideae, uniseptatae, exappendiculatae.

Lectotypus in foliis vivis *Caryae illinoensi*, "DeWitt, Georgia," legit J.B. Demaree (BPI).

Perithecia developing in irregular brown spots in living leaves, numerous in spots, perithecia globose depressed, 160-300 μm diam, 120-200 μm high, beaks erumpent hypophyllously, short to elongate, 110-300 μm long, 30 μm wide near base, enlarged below apex to 40 μm wide; peridium brown, 8-11 μm wide. Asci 40-55 × 9-13 μm, ellipsoid, often tapering below to slender stipe, (one) two (three, four) spored. Ascospores 23-39(-51) × 4.5-9(-10.5) μm, hyaline, ellipsoid, ends rounded, straight to inequilateral, septum median or variable, not appendaged.

Parasitic, maturing in spots in living leaves of *Carya illinoensis* (Wang.) K. Koch. southeastern North America.

Material examined: *Georgia:* DeWitt, J.B. Demaree (lectotype, BPI); Albany, 1932, 1933, Aug 1934, J.B. Demaree; Thomasville, Oct 1930 (2 collections), J.B. Demaree; *Florida*: Gainesville, 6 Feb 1917, J. Matz (all BPI).

Demaree and Cole (**1936**) described the symptoms of the fungus on pecan leaves. In culture perithecia matured in 20 days on corn meal agar. A Latin diagnosis was not given in the original publication, and one is provided here to validate the species.

Gnomonia section **Fusispora** Barr, sectio nov.

Ascosporae latae, fusoidae, in extrema acuta gradatim decrescentiae, septum medium, appendiculae latae setaceae vel obtusae; rostra lata. Species typica: *Gnomonia dryadis* Auerswald.

Ascospores broad, fusoid, tapering to pointed ends, usually inequilateral or curved, septum median, appendages broadly setose or blunt, overlapping biseriate in the ascus. Beaks broad.

Type species: *Gnomonia dryadis* Auerswald.

Gnomonia appendiculata Kar & Maity, Canad. J. Botany **48**: 1297. 1970, belongs in this series. The fungus, on canes of *Rubus* sp. from India, is distinguished especially by the simple or branched dark outgrowths extending the length of the broad beak. Ascospores are 12-15 × 4-5 μm, inequilateral to slightly curved, tapering to pointed ends, and appendages are relatively short and wide, almost triangular.

Gnomonia sanguisorbae (Rehm) Müller in Müller and von Arx, Beitr. Krypt. Fl. Schweiz **11**(2): 744. 1962, has slightly narrower fusoid ascospores, 7-16 × 2-2.5 μm with ± triangular appendages, according to Bolay (**1972**).

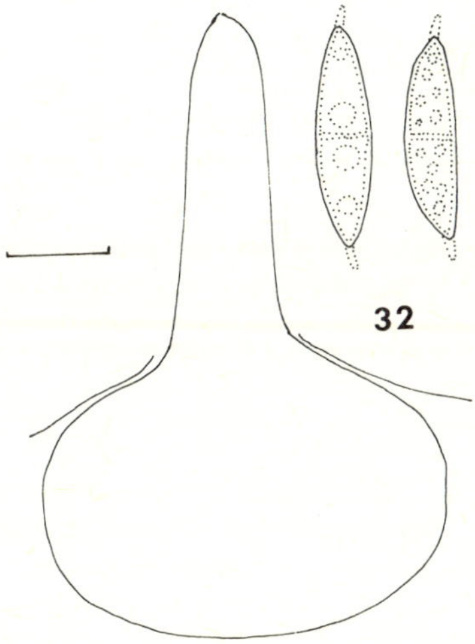

32

FIG. 32. Species of *Gnomonia* section *Fusispora*. *G. dryadis*, outline of perithecium, ascospores. Scale =100 μ or 10 μm.

23. **Gnomonia dryadis** Auersw., Myol. europ. **5/6**: 26. 1869. FIG. 32

Perithecia 385-540 μm diam, 277-308 μm high, depressed globose, beak straight to slightly curved, 260-385 μm long, 120-185 μm wide near base. Asci 80-96 × 13-16 μm. Ascospores 16-24 × 5-6.5 μm, fusoid, broadest in the middle and tapering to pointed ends, hyaline, straight to inequilateral or slightly curved, septum median, appendages blunt, 1.5-2.5 μm long, overlapping uniseriate to biseriate in the ascus.

In overwintered stalks of *Dryas* spp., Europe; *D. integrifolia* Vahl, North America.

Material examined: *North America: Newfoundland:* Hebron, Labrador, 24 Jul 1954, R.T. Wilce (DAOM, MASS).

Gnomonia section **Latispora** Barr, sectio nov.

Ascosporae ellipsoidae, in extrema obtusae gradatim decrescentiae, septum medium, ex appendiculae vel appendiculae setaceae; rostra angusta vel lata. Species typica: *Gnomonia intermedia* Rehm.

Ascospores ellipsoid, tapering to rounded ends, septum median, appendages setose when formed, overlapping obliquely in the ascus. Beaks narrow.

Type species: *Gnomonia intermedia* Rehm.

24. **Gnomonia intermedia** Rehm, Ann. Mycol. **6**: 489. 1908. var. **intermedia** FIG. 33

Perithecia 200-330 μm diam, 190-250 μm high, beaks 200-500 μm long, 32-50 μm wide near base, tapering to 24-40 μm near apex; peridium 7.5-12 μm wide. Asci 27-33 × 6-9 μm. Ascospores 9-12 × 2-3 μm, septum median, not appendaged, oblique or overlapping in the ascus.

On overwintered leaves of *Betula* spp., Europe, North America.

Material examined: *Europe:* Rehm, Ascomycetes 1794 (MICH, NY, isotype). *North America: Quebec:* Barr 1900, 1901, 1974, 2039, (DAOM, MASS); *Ontario:* Old Woman Bay, 23 Jun 1961, J.H. Mirza (TRTC); *Maine:* Barr, 3323, 3358, 3543 (MASS); *New Hampshire:* Barr 3870 (MASS); *Massachusetts:* Barr 6042 (MASS).

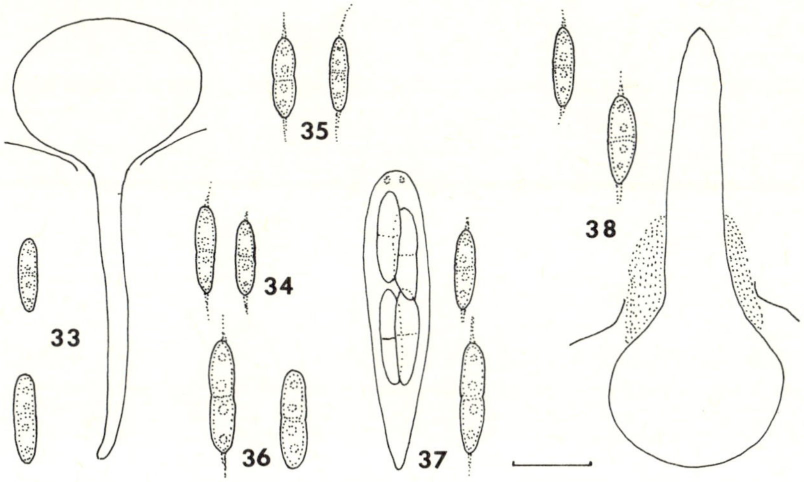

FIGS. 33-38. Species of *Gnomonia* section *Latispora*. 33. *G. intermedia* var. *intermedia*, outline of perithecium, ascospores. 34. *G. intermedia* var. *alni*, ascospores. 35. *G. rubicola*, ascospores. 36. *G. sibbaldiae*, ascospores. 37. *G. rubi*, ascus, ascospores. 38. *G. pulcherrima*, outline of perithecium, ascospores. Scale =100 μm or 10 μm.

Gnomonia setacea f. *betulae* Rehm, Ann. Mycol. **7**: 405. 1909, could be a synonym of *Gnomonia intermedia* rather than a form of *G. setacea*, according to the description. Klebahn (**1918**) obtained perithecia in culture from ascospores, as did Morgan-Jones (**1953**, **1959**). A conidial state in also produced in culture. Conidia are oblong-cylindroid, hyaline, one celled, 7-8.5 × 1-2.5 μm, according to Klebahn. In cultures from North American specimens, smaller conidia, 4-6 × 1-1.5 μm, were produced from short phialides lining pycnidia.

25. **Gnomonia intermedia** var. **alni** Barr, var. nov. FIG. 34

Perithecia 150-385 μm diametro, 165-275 μm alto, rostra (110-)250-600 μm longa, 25-45 μm lata. Asci 24-35(-40) × 5-9 μm. Ascosporae 9-12 × (1.5-)2-3 μm, ellipsoideae, hyalinae, septum medium, setoso-appendiculatae.

In foliis anni praeteriti *Alni rugosae* var. *americanae*. Holotypus "Basin Pond, Baxter State Park, Maine, 5 Jul 1962," a M.E. Barr n. 3322 lecto (MASS).

Perithecia 150-385 μm diam, 165-275 μm high, beaks (110-)250-600 μm long, 25-45 μm wide near base. Asci 24-35(-40) × 5-9 μm. Ascospores 9-12 × (1.5-)2-3 μm, ellipsoid, septum median, appendages setose, up to 5 μm long, oblique in the ascus.

In overwintered leaves of *Alnus rugosa* (Du Roi) Spreng. var. *americana* (Regel) Fern., North America.

Material examined: *Manitoba:* WINF(M) 2257a, 2988b (WINF); *British Columbia:* Barr 5582, 5776 (MASS); *Maine:* Barr 3322 (holotype, MASS); *Massachusetts:* Barr 5597 (MASS).

This species is similar to *G. intermedia* var. *intermedia* except for the setose appendages on the ascospores, and the host plant. Vleugel's *G. setacea* f. *alni* (Svensk Bot. Tidskr. **5**: 330. 1911) could be identical with the North American collections: the ascospores are said to be up to 4 μm wide when mature. However, Klebahn's (**1918**) detailed description and illustrations are of *G. setacea.*

26. **Gnomonia ribicola** Barr, sp. nov. FIG. 35

Perithecia 250-540 μm diametro, 200-500 μm alto, rostra (250-)500-900 μm longa, 30-53 (-65) μm lata. Asci 24-42(-50) × (4.5-)6.5-10 μm. Ascosporae (6.5-)8.5-12(-13) × 2-3.5 (-4) μm, hyalinae, septum medium, ellipsoideae, setoso-appendiculatae.

In foliis anni praeteriti et ramulis *Ribis* spp. Holotypus "Basin Pond, Baxter State Park, Maine, 5 Jul 1962," a M.E. Barr n. 3308 lecto (MASS).

Perithecia 250-540 μm diam, 200-500 μm high, beaks (250-)500-900 μm long, 30-53(-65) μm wide near base, rarely lacking. Asci 24-42 (-50) × (4.5-)6.5-10 μm. Ascospores (6.5-)8.5-12(-13) × 2-3.5(-4) μm, ellipsoid, septum median, appendages setose, delicate, 7-11 μm long, occasionally lacking, oblique or overlapping in the ascus.

In leaves and twigs of *Ribes* spp., North America.

Material examined: *Quebec:* Barr 1925a; *British Columbia:* Barr 634c; *Maine:* Barr 3308, holotype; *Oregon:* Barr 6286 (all MASS).

Gnomonia ribicola is similar to the two varieties of *G. intermedia* in most respects. The perithecia are larger and ascospores wider. One collection (Barr 634) was found in pith of broken branches and the perithecia were unbeaked, with only a periphysate ostiolar pore, apparently in response to the enclosed habitat.

Only three poorly known species have been described from *Ribes*. *Gnomonia grossulariae* (Fries) Sacc. (Syll. Fung. **1**: 570. 1882) (*Sphaeria grossulariae* Fries, Syst. Mycol. **2**: 521. 1823) was described with perithecia collapsing as in *Sphaeria setacea* but without a prominent beak (Fries, **1823**). Saccardo (**1882**) and Winter (**1887**) regarded it as a doubtful species, and Fuckel (1870) transferred it to *Pleospora*. Klebahn (**1918**), Shear (**1943**), and Wehmeyer (**1961**) have relegated the name to synonymy under *Pleospora herbarum* (Fries) Rabenh., after study of Fries's Scler. suec. exs. n. 57. The second possible species was *Sphaeria circinata* Fuckel (Symb. Mycol. p. 111. 1870), which Saccardo transferred to *Gnomoniella* in Sylloge Fungorum **1**: 416. 1882 with doubt, suggesting that it might be a species of *Physalospora*. The description suggests *Physalospora*, certainly not *Gnomonia* or *Gnomoniella*. Winter (**1887**) had the fungus in *Didymosphaeria*. *Sphaeria curva* Wallr. (*Gnomonia curva* (Wallr.) Fckl., Symb. Mycol. p. 123. 1870) was noted as a rarely found species and the extremely brief description, without observations on asci or ascospores, makes it a doubtful name.

27. **Gnomonia sibbaldiae** Petrak, Krypt. Forsch. Bayer. Bot. Ges. **2**: 162. 1931. FIG. 36

Perithecia 195-350 μm diam, 220-270 μm high, beaks often curved, 200-285 μm long, 100-120 μm wide near base; peridium 15-16.5 μm wide. Asci 50-60 × 7.5-11 μm, apical annulus small. Ascospores (9-)12-16.5 × 3-4.5 μm, ellipsoid, septum median, appendages setose, 3-3.5 μm long when present, biseriate to triseriate in the ascus.

On leaves of *Sibbaldia, Sieversia, Dryas* spp., Europe, North America.

Material examined: *Europe: Sibbaldia procumbens* L., Tirol: Arlberggebiet: Mulfontal, 19 Aug 1928, A. Ade (NY, type). *North America: Northwest Territories: Dryas integrifolia* Vahl., head of Clyde Inlet, Baffin Island, 24 Jun 1950, P. Dansereau (MASS).

Bolay (**1972**) noted that ascospores of *G. sibbaldiae* brown after discharge. He was unable to obtain cultures because the ascospores failed to germinate.

28. **Gnomonia rubi** (Rehm) Winter in Rabenh., Kryptogamenfl. **1**(2): 586. 1887; Hedwigia **26**: 62. 1887. FIG. 37

Gnomonia tetraspora Winter var. *rubi* Rehm, Ascomyceten no. 596; Hedwigia **20**: 53. 1881.

Perithecia 165-280 μm diam, 135-250 μm high, scattered to ±grouped, beaks 130-470 μm high, 80-100 μm wide near base; peridium 10-22 μm wide. Asci 33.5-44 × 5-8 μm, four spored. Ascospores 10-15.5 × 2.5-3.5 μm, ellipsoid, septum median, appendages setose, 1.5-6 μm long.

In leaves and canes of *Rubus* spp., Europe, North America.

Material examined: *North America: Nova Scotia:* Upper Brookside, Colchester Co., 21 Jul 1929, L.E. Wehmeyer (MICH); *Ontario;* TRTC 9039 (TRTC); *Vermont:* Greensboro Bend, 29 Aug 1924, L.O. Overholts & C.R. Orton (TRTC); *Massachusetts:* Barr 6039 (MASS); *Oregon:* Corvallis, no date, S.M. Zeller; Springbrook, 21 Mar 1924, S.M. Zeller; Corvallis, 21 Apr 1924, S.M. Zeller (NY).

I am doubtful whether *G. rubi* is identical with or is a four-spored form of *G. rostellata* (Fr.) Wehm. (**1942**). In Wehmeyer's collection from Nova Scotia labelled as *G. rostellata*, the asci have disintegrated so that their ascospore number could not be observed. Perithecia are similar, ascospores are within the same size range, have similar shape, septation, and appendages to those described as *G. rostellata*. However, I have seen no other collections of *G. rostellata* with intact asci.

29. **Gnomonia pulcherrima** Seaver & Waterston, Mycologia **32**: 402. 1940. FIG. 38

Perithecia 200-270 μm diam, globose, not collapsing, in petioles and main veins, beaks 240-560 μm long, 130-180 μm wide at base, tapering to 70-90 μm near apex, surrounded by collar of yellowish granular material; peridium 10-12 μm wide. Asci 33-44 × 7.5-13 μm. Ascospores 9-12 × 2.5-4 μm, ellipsoid, septum median, appendages setose, 2-3 μm long.

In petioles and main veins of leaves of *Coccoloba uvifera* (L.) Jacq. Bermuda.

Material examined: *Bermuda:* Bermuda Fungi 82 (type), 325, 338b, 340 (NY).

Although the illustration with the original description shows three-septate ascospores, the specimens and the description indicate that they are one septate.

Gnomonia section **Clava** Barr, sectio nov.

Ascosporae latae ellipsoidae, in extrema obtusa gradatim dcrescentiae, septum submedium, exappendiculae vel appendiculae setaceae; rostra plerumque clavata, angustati et apicem versus dilatata. Species typica: *Gnomoni myricae* Cooke & Ellis.

Ascospores broadly ellipsoid, ends rounded, septum submedian, appendages setose when formed. Beaks narrow.

Type species: *Gnomonia myricae* Cooke & Ellis.

30. **Gnomonia myricae** Cooke & Ellis, Grevillea **6**: 17. 1877. FIG. 39

Apiognomonia myricae (Cooke & Ellis) Dennis, Kew Bull **32**: 118. 1977.

Perithecia 230-300 μm diam, ca. 150 μm high, amphigenous but mostly hypophyllous, beaks 140-440 μm long, 60-75 μm wide near base, flared to 105 μm wide below apex or tapering at times, usually curved; peridium 10-12 μm wide. Asci (25-)22-47 × 6-10.5 μm. Ascospores (7.5-) 9-12 × 2.5-4(-5) μm, septum submedian or nearly median at times, appendages 1-1.5 μm long.

In overwintered leaves of *Myrica* spp., North America.

Material examined: *New Jersey:* Newfield, May 1877, J.B. Ellis (isotype, BPI); Ellis & Ev. N.A.F. 2137 (BPI, MASS, MICH); *Georgia:* 4 miles west of Harlem, Columbia Co., 16 Mar 1939, J.H. Miller (GA 6606; FLAS 25515, GA 6607); *California:* H.E. Parks 4164 =Calif. Fungi 1033 (GA, MICH, TRTC).

31. **Gnomonia gei-montani** Ranojevic, Ann. Mycol. **8**: 362. 1910. FIG. 40

Perithecia 220-250 μm diam, globose, immersed, scattered, beaks erumpent through lower leaf surface, 200 μm long, ca. 80 μm wide near base, tapering to 40 μm wide near apex. Asci 30-44 × 6.5-7.5 μm, oblong. Ascospores 9-11 × 2.5-3.5 μm, ellipsoid, septum slightly submedian, appendages setose.

In overwintered leaves of *Geum* spp., Europe, North America.

Material examined: *New Hampshire:* Barr 4095 (MASS).

This collection agrees in most respects with the type description and with that provided by Bolay (**1972**) for *G. gei-montani*. However, the

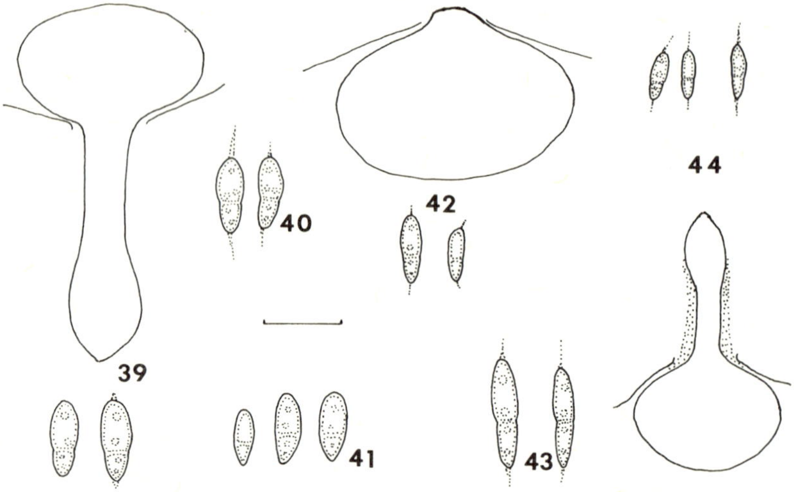

FIGS. 39-44. Species of *Gnomonia* section *Clava*. 39. *G. myricae*, outline of perithecium, ascospores. 40. *G. gei-montani*, ascospores. 41. *G. clavulata*, ascospores. 42. *G. rauii*, outline of perithecium, ascospores. 43. *G. misella*, ascospores. 44. *G. dalibardae*, outline of perithecium, ascospores. Scale =100 μm or 10 μm.

perithecial beaks are slightly broader, as are the ascospores (7.5-15 × 1.5-3.2 μm in the type), so perhaps the North American material should be recognized as a separate subspecies. This is a subalpine species, with the type collected at ca. 2000 m. in Serbia and the North American material from the Alpine Garden area of Mt. Washington.

32. **Gnomonia clavulata** Ellis, Amer. Nat. **17**: 318. 1883. FIG. 41

Didymella clavulata (Ellis) Sacc., Syll. Fung. **9**: 666. 1891 (as *Didymella clavulata* Ellis).

Cercidospora clavulata (Ellis) Kuntze, Rev. Gen. Pl. **3**(2): 453. 1898.

Perithecia 160-200 μm diam, 140-170 μm high, depressed globose, scattered singly or grouped in leaf tissues, beaks erumpent through lower leaf surface, 150-220 μm long, 20-30 μm diam near base, tapering or enlarged below apex to 36-40 μm diam; peridium 9-15 μm wide. Asci 25-38.5 × 6-9 μm. Ascospores 7.5-10.5 × 3-4 μm, hyaline, yellowish in mass, ellipsoid, narrowed more to base than apex, straight, inequilateral or slightly curved, septum submedian, overlapping biseriate in the ascus.

In overwintered leaves of *Quercus nigra* L., *Carya* sp., eastern North America.

Material examined: *New Jersey*: Newfield, May 1882, J.B. Ellis (type, NY); Ellis & Ev. N.A.F. 1685 (BPI, MASS, MICH); N.A.F. 3429 (MASS, as and with *Gnomonia setacea* var. *macrospora*).

33. **Gnomonia rauii** (Ell. & Ev.) Petrak, Sydowia **15**: 207. (1961) 1962. FIG. 42

Sphaeria (Didymella) rauii Ell. & Ev., Bull. Torrey Bot. Club **10**: 90. 1883.

Didymella rauii (Ell. & Ev.) Ellis, N.A.F. n. 1336, 1885; Berl. & Vogl. in Sacc., Syll. Fung. Addit. **1-4**: 96. 1886; Syll. Fung. **9**: 668. 1891.

Endophlaea (Didymella) rauii (Ell. & Ev.) Cooke, Grevillea **17**: 89. 1889.

Cercidospora rauii (Ell. & Ev.) Kuntze, Rev. Gen. Pl. **3**(2): 454. 1898 (as "rani").

Perithecia 275-400 μm diam, 200-250 μm high, depressed globose, immersed, scattered, beaks short papillate or more elongate and broad, 80-100 μm long, 100-150 μm wide; peridium 16-22 μm wide. Asci (21-) 24.5-38 × 4.5-7.5 μm, ellipsoid. Ascospores 7-10 × 1.5-2.5 μm, hyaline, ellipsoid, ends rounded, straight to curved, septum submedian, appendages setose, 1 μm long, overlapping biseriate in the ascus.

In dead twigs of cultivated *Rosa* sp., eastern North America.

Material examined: *Pennsylvania:* Bethlehem, 20 May 1883, Rau (type, NY) (as *Diaporthe rauii*); Ellis N.A..F. 1336 (MASS, MICH).

Wehmeyer (**1933**) examined the type specimen which was filed under the unpublished name of *Diaporthe rauii* Ell. & Ev. He noted that it belonged in *Gnomonia* and provided a brief description, although he did not officially transfer the species to *Gnomonia.*

34. **Gnomonia misella** Niessl, Oesterr. Bot. Zeitschr. p. 48. 1875; Hedwigia **14**: 146. 1876. FIG. 43

Cryptoderis misella (Niessl) Sacc., Syll. Fung. **2**: 230. 1883.

Gnomoniopsis chamaemori var. *misella* (Niessl) Berl., Icon. Fung. **1**: 94. 1892.

Perithecia 245-330 μm diam, 220-245 μm high, depressed globose, beak short, 190 μm long, 110 μm wide and enlarged below apex. Asci 45-50 × 6-7 μm, four spored. Ascospores 11-16 × 2.5-3.5 μm, ellipsoid, appendages 3-5 μm long.

In old stalks of *Hypericum* and *Oenothera* spp., Europe; *Oenothera biennis* L., North America.

Material examined: *Ontario:* Univ. of Toronto Forest, Haliburton Co., 12 Sep 1968, D. Malloch 645 (TRTC, MASS).

Niessl (**1875**) noted particularly the immersed perithecia with short conic beaks in his original description; he described the ascospores in the four-spored asci as three septate and four guttulate, with setose appendages. Only a single septum was found in ascospores of the North American collection.

35. **Gnomonia dalibardae** Barr, sp. nov. FIG. 44

Perithecia 140-200 μm diametro, 120-140 μm alto, globosa depressa, rostra 140-200 μm longa, fundi 35 μm et apices 50-70 μm diametro, collum flavidum rostri cingens. Asci 26.5-33 × 5.5 μm. Ascosporae 6.5-7.5 × 2-2.5 μm, hyalinae, latae ellipsoidae, in extremo obtusa gradatim decrescentiae, septum submedium, appendiculae setaceae.

Holotypus in foliis anni praeteriti *Dalibardae repentis,* "Cedar Lake, Penobscot Co., Maine, 24 Aug 1962," a M.E. Barr n. 3689 lecto (MASS).

Perithecia 140-200 μm diam, 120-140 μm high, depressed globose, beaks 140-200 μm long, narrow at base, ±35 μm, enlarged below the apex to 50-60 μm, surrounded by collar of yellowish material; peridium 9-11 μm wide. Asci 26.5-33 × 5.5 μm. Ascospores 6.5-7.5 × 2-2.5 μm, ellipsoid, septum submedian, ends rounded, appendages short setose.

In overwintered leaves of *Dalibarda repens* L., eastern North America.

Material examined: *Maine:* Barr 3582, 3689 holotype (MASS).

This species is similar in many respects to some variants of *G. comari.* The ascospores are relatively wider and bear setose appendages; the yellowish collar surrounding the beak is another distinguishing feature. *Gnomonia waldsteiniae* has a whitish collar and narrower ascospores without appendages.

4. **Ophiognomonia** (Sacc.) Sacc., Syll. Fung. **14**: 613. 1899.

Gnomoniella subgenus *Ophiognomonia* Sacc., Syll. Fung. **1**: 419. 1882.

Stromatic tissues not formed; perithecia immersed, upright, beaks central rarely eccentric, elongate, narrow. Asci elongate cylindroid. Ascospores hyaline, elongate, narrowly clavoid, enlarged slightly and usually septate in upper third, tapering to base, in a fascicle in the ascus.

Conidial state not known.

Type and single North American species: *Ophiognomonia melanostyla* (DC. ex Fries) Sacc.

When Saccardo (**1882**) created the subgenus *Ophiognomonia*, he included only *G. melanostyla* in it. Von Höhnel (**1917**) suggested as lectotype species *O. procumbens* (Fckl.) v. Höhnel, in **1918** listed both *O. melanostyla* and *O. procumbens*, and in **1919** stated that *Ophiognomonia*, with upright perithecia and central beak, was typified by *O. melanostyla*, not *O. procumbens*, which has oblique perithecia and a lateral beak.

1. **Ophiognomonia melanostyla** (DC. ex Fries) Sacc., Syll. Fung. **14**: 613. 1899; in Berlese, Icon. Fung. **2**: 146. 1899. FIG. 45

Sphaeria melanostyla DC. ex Fries, Syst. Mycol. **2**: 517. 1823.

Gnomonia melanostyla (DC. ex Fries) Auersw., Leipzig Tausch-Verein. p. 4. 1866.

Gnomoniella melanostyla (DC. ex Fries) Sacc., Syll. Fung. **1**: 419. 1882.

Cryptoderis melanostyla (DC. ex Fries) Winter in Rabenh., Kryptogamenfl. **2**: 592. 1887.

Pleuroceras melanostyla (DC. ex Fries) v. Höhnel, Ber. Deutsch Bot. Ges. **35**: 637. 1917.

Perithecia 200-350 μm diam, 200-280 μm high; beaks 300-770 μm long, 28-40 μm wide near base. Asci 45-75 × 3-5 μm. Ascospores 25-48 × 0.5-1.5 μm, septum supramedian, 1-1.5 μm wide above septum, 0.5-1 μm wide below septum.

In leaves and petioles of *Tilia* spp., Europe, North America.

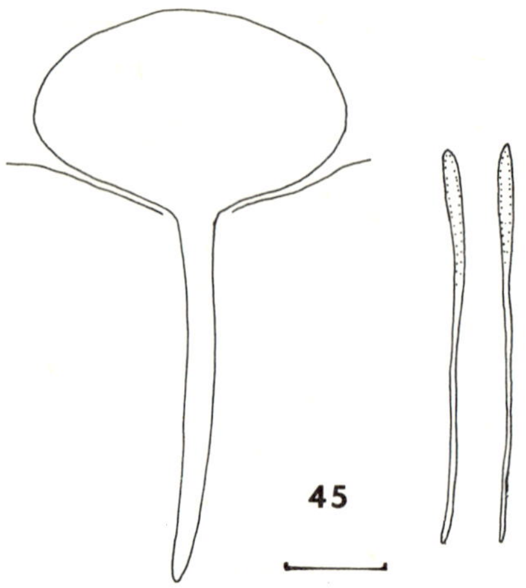

FIG. 45. *Ophiognomonia melanostyla,* outline of perithecium, ascospores. Scale =100 μm or 10 μm.

Material examined: *Europe:* Herb. Barbey-Boissier 83 (DAOM, TRTC); Sydow, Mycotheca germ. 385 (TRTC); Thümen, Myc. univ. 265 (BPI). *North America: Ontario:* Nashville, York Co., 16 May 1954, R.F. Cain (TRTC); London, June 1894, J. Dearness (BPI, NY); Granton, 30 Apr 1894, J. Dearness (NY); *New York:* Helderberg Mts., May, C.H. Peck (NYS).

Klebahn (**1918**) studied the process of ascospore germination. The upper cell of the ascospore swelled to 3-3.5 μm and often became septate while the narrow lower cell did not increase in size. Germ tubes developed from the one or two upper swollen cells. No conidial state was formed, but perithecia were produced in agar culture and on sterilized *Tilia* leaves.

I-2. **Gnomoniaceae** subfamily **Stegophoroideae** Barr, subfam. nov.

Perithecia singulara vel aggregata, stromata contexta prosenchymata, saepe instar clypei vel zonae nigrifactae vel disci stromatici. Genus typicus: *Stegophora* Sydow & Sydow (two tribes).

I-2a. Tribe **Stegophoreae**

Perithecia usually single in substrate, clypeus present; in leaves and petioles of deciduous trees and shrubs, or in culms of monocotyledonous plants.

Key to Genera of *Stegophoreae*

1. Ascospores nonseptate, ellipsoid or fusoid .... 5. *Sphaerognomonia*
1. Ascospores septate or elongate 2
    2. Ascospores apiosporous, ellipsoid .... 6. *Stegophora*
    2. Ascospores elongate .... 7. *Linocarpon*

Extralimital genera belonging in this tribe include *Bagcheea, Diplacella*, and *Uleoporthe*. The type species of *Bagcheea, B. castaneae* E. Müller & Menon (=*B. albomaculans* (Fukui) Hino & Katumoto), from Asia on living leaves of species of *Castanopsis*, has short papillate perithecia, thickly scattered in leaf spots. The one-celled ascospores, which have a refractive central area and granular ends, are reminiscent of those found in species of the Physciaceae of the Lecanorales. Recent descriptions and illustrations are in Müller and Menon (**1954**) and Katumoto (**1965**) as well as in Kobayashi (**1970**).

Diplacella Sydow (Ann. Mycol. **28**: 101. 1930) with the type species *D. paullinae* (Frag. & Cif.) Sydow (=*D. mararyensis* (P. Henn.) von Arx & Müller, **1954**) is an extralimital fungus apparently related to *Sphaerognomonia*. The sole species is a leaf parasite on *Paullinia, Serjana*, and other species of the Sapindaceae in Central and South America. It was fully described by von Arx and Müller (**1954**) and was illustrated by von Arx (**1951**).

Uleoporthe Petrak (Ann. Mycol. **39**: 279. 1941), typified by *U. orbiculata* (H. Syd.) Petrak, in leaves of a species of Myrsinaceae, South America, was included as a synonym of *Phylloporthe* by Müller and von Arx (**1962**). However, the stromatic tissues of *Uleoporthe* are prosenchymatous, whereas those of *Phylloporthe vernoniae* Syd. are pseudoparenchymatous. The ascospores of *U. orbiculata* are hyaline with a medium septum.

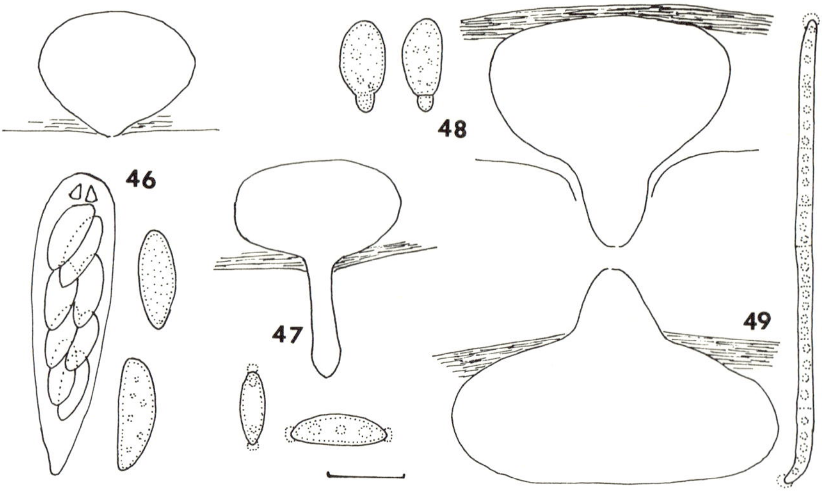

FIGS. 46-49. FIGS. 46-47. Species of *Sphaerognomonia*. 46. *S. carpinea*, outline of perithecium, ascus, ascospores. 47. *S. papillostoma*, outline of perithecium, ascospores. FIG. 48. *Stegophora ulmea*, outline of perithecium, ascospores. FIG. 49. *Linocarpon pandani*, outline of perithecium, ascospore. Horizontal lines indicate prosenchymatous clypeal tissues. Scale =100 μm or 10 μm.

5. **Sphaerognomonia** Potebnia, Ann. Mycol. **8**: 53. 1910.

Apiosporopsis (Trav.) Mariani, Atti Soc. Ital. Sci. Nat. **50**: 165. 1911.

Stromatic tissues prosenchymatous, as slight or well-developed clypeus over perithecial apex; perithecia upright, immersed, depressed globose, beaks as apical pore only, short papillate, or elongate and broad. Asci ellipsoid, (four) eight spored. Ascospores one celled, thin walled, hyaline, relatively small, ellipsoid or fusoid, often inequilateral or slightly curved.

Conidial state formed in culture as stroma containing locule; conidiogenous cells phialidic; conidia one celled, hyaline, oblong or cylindroid to allantoid.

In overwintered leaves or petioles of hardwood trees and shrubs.

Type species: *Sphaerognomonia carpinea* (Fries) Potebnia.

Sphaerognomonia carpinea was the original single species of the genus. The absence of a beak and thin-walled, one-celled ascospores were the diagnostic characters of the genus. A short beak is found in a number of collections, however, and the basic character of the genus is considered to be the thin-walled, one-celled ascospores, in conjunction with the formation of a thin clypeus surrounding the beak (or apical region), visible in thin sections of perithecia. The other species accommodated here has relatively broad beaks of varying length, in addition to a well-developed clypeus.

Miller and Thompson (**1940**) transferred *Anisostomula polystigma* (Ell. & Ev.) v. Höhnel to *Sphaerognomonia.* However, this species differs in a number of features — in apiospores as well as in centrum characteristics — and it belongs in *Pseudomassaria* in the Physosporellaceae. *Hyponectria cookeana* (Auersw.) Barr on *Quercus* leaves may be confused with *S. carpinea.* Both species have one-celled ascospores, which in *H. cookeana* are usually smaller, 8-10.5 × 3-4 μm, in asci which have an amyloid apical annulus. Delicate paraphyses are present, and the peridium differs in structure from that of *S. carpinea.* Kobayashi (**1970**) described *Sphaerognomonia haraeana* on leaves of *Cryptomeria,* and suggested that it might be intermediate between *Sphaerognomonia* and *Gnomoniella.* The so-called clypeus appears to be a white collar surrounding the upper part of perithecium and base of beak, rather than a hyphal prosenchymatous tissue, and *S. haraeana* is probably a good species of *Gnomoniella.*

Key to Species of *Sphaerognomonia*

1. Perithecia without beak, or beak short; ascospores 10-15(-18) × 3.5-6.5(-9) μm, not appendaged .... 1. *S. carpinea*
1. Perithecia with well-developed, broad beak; ascospores 9-13 × 2.5-4(-4.5) μm, with blunt appendages .... 2. *S. papillostoma*

1. **Sphaerognomonia carpinea** (Fries) Potebnia, Ann. Mycol. **8**: 54. 1910.
FIG. 46

Sphaeria carpinea Fries, Syst. Mycol. **2**: 523. 1823.

Ascospora carpinea (Fries) Fries, Summa Veget. Scand. p. 425. 1846.

Sphaerella carpinea (Fries) Auersw. in Rabenh., Mycol. Europ. **5/6**: 2. 1869.

Laestadia carpinea (Fries) Sacc., Syll. Fung. **1**: 426. 1882.

Guignardia carpinea (Fries) Schroeter in Cohn, Die Pilze Schlesiens **2**: 330. 1893.

Carlia carpinea (Fries) Kuntze & Bon. in O. Kuntze, Rev. Gen. Pl. **2**(3): 846. 1898.

Apiosporopsis carpinea (Fries) Mariani, Atti Soc. Ital. Sci. Nat. **50**: 165. 1911.

Gnomonia carpinea (Fries) Klebahn, Askomyzeten, p. 274. 1918.

Laestadia orientalis Ell. & Ev., Proc. Acad. Nat. Sci. Phila. for **1890**: 230. 1891.

Laestadia castanicola Ell. & Ev., North Amer. Pyreno. p. 259. 1892.

Perithecia 112-250 μm diam, (80-)140-170 μm high, globose or depressed, immersed, usually hypophyllous, beak absent and apex rounded with plane pore or short papillate or conic, 70-140 μm wide, 50-70(-200) μm high; clypeus slight, prosenchymatous, peridium 10-20 μm wide. Asci (37-)40-63(-78) × (7.5-)9-14.5 μm, apical annulus shallow. Ascospores 10-15(-18) × 3.5-6.5(-9) μm, ellipsoid, ovoid, or fusoid, straight or often inequilateral, overlapping biseriate or obliquely uniseriate in the ascus.

Conidial state: Potebnia (**1910**) described the conidia from culture, produced on scattered or grouped hyphae, ovoid, 12-14(-17) × 4.5-6 μm. He noted the similarity of conidium production in *S. carpinea* and *Gnomonia veneta* (=*Apiognomonia errabunda*), and postulated that *Gloeosporium robergei* Desm. (*Monostichella robergei* (Desm.) v. Höhnel) was the conidial state on living leaves. Klebahn (**1918**) also obtained conidia from hyphae in culture. He described and illustrated pycnidium-like structures which developed after more mycelium had formed and which also produced conidia, 9-16 × 5-6.5 μm. Klebahn disagreed with Potebnia's association of *G. robergei* as the conidial state; he associated a species of *Leptothyrium* on *Carpinus* leaves with the perithecial fungus produced on overwintered leaves. Klebahn was unsuccessful in infection experiments with the ascospores on young *Carpinus* leaves, and postulated that only aging leaves could be infected.

On overwintered leaves of numerous hardwood hosts: species of *Acer, Alnus, Betula, Carpinus, Castanea, Corylus, Fraxinus, Ostrya, Quercus,* from Europe and North America.

Material examined: *Europe:* Krieger, Fungi saxonici 1467 (MICH); Sydow, Mycotheca germ. 1333, 2698, 3119 (MICH). *North America: Ontario:* London, Apr 1884, J. Dearness (BPI); *Massachusetts:* Barr 2552, 2560, 5080, 5081, 5623 (MASS); *Rhode Island:* Barr 4750 (MASS); *New Jersey:* Newfield, 4 Apr 1891, J.B. Ellis (type of *Laestadia castanicola*, NY); *Virginia:* Radnor, May 1920, C.L. Shear (BPI); *West Virginia:* Ell. & Ev. Fungi Col. 1326 (MASS, MICH); Fayette Co., 20

Apr 1898, L.W. Nuttall (MICH, BPI); *Georgia:* GA 8604-8620, 6689 (GA); Barr 6178, 6193 (MASS); *Louisiana:* Lafayette, 21 Mar 1888, A.B. Langlois (Flora Ludoviciana 1664, type of *Laestadia orientalis*, NY).

2. **Sphaerognomonia papillostoma** (Dearn. & House) Barr, comb. nov. FIG. 47

Gnomonia papillostoma Dearness & House, Bull. New York State Mus. **243-244**: 74. 1923.

Perithecia 200-300 μm diam, 130-200 μm high, depressed globose, immersed, separate or grouped, bases visible through upper leaf epidermis, beak erumpent through small blackened clypeus, central to eccentric, 50-220 μm long, 50-70 μm wide, even or enlarged below pallid apex; peridium 20-30 μm wide. Asci 27.5-44 × 7.5-11 μm, (four) eight spored. Ascospores 9-13 × 2.5-4(-4.5) μm, hyaline, narrowly ellipsoid fusoid, one celled, straight or inequilateral, wall thin, appendages blunt, 1-2 μm long, 2-3 μm wide, overlapping biseriate or oblique in the ascus.

Conidial state: In culture from ascospores, loculate stromata develop in mycelial mat, 2.5-8 mm diam, 4-7 mm high, rounded or irregular in shape, "beaked" at times, containing one or several locules; conidiogenous cells lining locules, short to elongate, simple or branched, phialidic; conidia 4-5 × 1-1.5 μm, oblong or cylindroid to allantoid, hyaline, one celled.

In overwintered leaves which usually remain attached to twigs of *Spiraea latifolia* (Ait.) Borkh., North America.

Material examined: *Maine:* Barr 5858 (MASS); *New Hampshire:* Barr 3820, 3902 (MASS); *Vermont:* Barr 4279 (MASS); *Massachusetts:* Barr 5003, 5587, 5764 (MASS); *New York:* Albany, 1 May 1920, H.D. House (type, NYS); 25 Apr 1922 (NYS).

A collection in BPI from Klamath, California, 22 Jul 1913, collected by E.P. Meinecke, has the same aspect as the eastern collections. The perithecia are empty of asci and ascospores, so that identification is uncertain.

6. **Stegophora** Sydow & Sydow, Ann. Mycol. **14**: 364. 1916.

Stromatic tissues prosenchymatous, as pallid hyphae in leaf tissues, forming blackened crustose clypeus in upper epidermal layer; perithecia upright, immersed in leaf tissues, several beneath blackened clypeus, beaks central, erumpent through lower leaf epidermis, short and broad. Asci ellipsoid. Ascospores hyaline, broadly ellipsoid, septum near base (apiosporous), appendages not formed.

Conidial state: *Cylindrosporella* (von Arx, **1957**; Morgan-Jones, **1958**) (=*Asteroma*, Sutton, **1977**): produced as acervuli in blackened clypeal tissues in living leaves; conidiogenous cells phialidic; conidia hyaline, fusoid ellipsoid, one celled.

In leaves of deciduous trees.

Type species: *Stegophora ulmea* (Schw. ex Fries) Sydow & Sydow.

A second species, *S. oharana* (Nisikado & Matsumoto) Petrak (Ann. Mycol. **38**: 367. 1940), occurs on leaves of *Ulmus* spp. in eastern Asia, and has been separated from *S. ulmea* by the formation of perithecia in living leaves in autumn and by larger ascospores, 12.5-16 $\times$ 4-6.5 μm. It too produces a *Cylindrosporella* conidial state. The position of perithecia and beaks — perithecia horizontal, often paired, beaks lateral and erumpent together — leads me to suggest the *Hypospilina* may be a better disposition of this species.

Müller and von Arx (**1962**, **1973**) considered *Stegophora* to be a synonym of *Lambro* Racib., but the type species of *Lambro* has pseudoparenchymatous stroma surrounding three to twelve perithecia, forming a dark outer crust around the hyaline interior. Kobayashi (**1970**) retained *Stegophora* as distinct from *Lambro* on the basis of differences in stromatic tissues. Sydow and Sydow (**1916**), when they erected the genus, included *Stegophora* in the family Stegasphaeriaceae, based upon the formation of a clypeus at the base of perithecia. In **1936** Sydow recognized that the clypeal character alone was not sufficient to serve as a family characteristic, and that disparate genera were thus grouped together. *Stegasphaeria* Sydow & Sydow, the other genus of the family, was included in the Mesnieraceae of the Pseudosphaeriales by Müller and von Arx (**1962**).

Miles (**1921**) described the disease symptoms, conidial state, and development of perithecia of *Gnomonia ulmea*. Pomerleau (**1938**) studied the biology of the fungus and gave detailed information on development. The release of ascospores and infection are dependent upon favorable temperature and moisture conditions in the spring; rain and humidity during summer months determine the spread of the disease by conidia.

1. **Stegophora ulmea** (Schw. ex Fries) Sydow & Sydow, Ann. Mycol. **14**: 364. 1916. FIG. 48

Sphaeria (Confertae) *ulmea* Schw. ex Fries, Syst. Mycol. **2**: 436. 1823.

Gnomonia ulmea (Schw. ex Fries) Thümen, Flora **36**: 178. 1878.

Dothidella ulmea (Schw. ex Fries) Ellis & Ev., North Amer. Pyreno. p. 608. 1892.

Lambro ulmea (Schw. ex Fries) Müller in Müller & von Arx, Beitr. Kryptogamenfl. Schweiz **11**(2): 777. 1862.

Stromatic tissues forming shining black ectostromata up to 1 cm diam, epiphyllous in epidermal layer of living leaves; perithecia upright, few grouped beneath stromatic tissues, immersed in leaf tissues, depressed globose, 200-385 μm diam, (150-)215-230 μm high, beak central, 82-100 μm long, 85-105 μm wide, erumpent hypophyllously; peridium 26-33 μm wide. Asci 42-51 × 9-12 μm, apical annulus broad. Ascospores (7-)10-13.5 × 3-6 μm, hyaline, ellipsoid ovoid, straight, ends rounded, septum near base, lower cell conoid, 1.5-3 μm wide and long, wall smooth, overlapping biseriate to triseriate in the ascus.

Conidial state: *Cylindrosporella ulmea* (Miles) von Arx, **1957** (*Gloeosporium ulmeum* Miles, **1921**): Acervuli subepidermal or subcuticular, 1-2 mm wide or larger, 52-65 μm high, covering shining black, composed of a single layer of irregular dark brown cells, basal layer brown, 10-15 μm thick; conidiogenous cells up to 10 μm long, phialidic; conidia 4-6(-10) × 1-2 μm, hyaline, fusoid ellipsoid or cylindroid, one celled, minute globule near each end.

Developing in living leaves and maturing by December or after overwintering in fallen leaves of *Ulmus* spp., especially *U. americana* L., widespread and common throughout North America.

Material examined (representative perithecial collections): *Ontario:* Economic Fungi 155a, b (DAOM, FLAS, MASS); Macoun, Canadian Cryptogams 20 (TRTC), 619 (DAOM); Dearness, Canadian Fungi 208, 209 (DAOM); *Vermont:* Barr 4490; *Massachusetts:* Barr 2548; *New York:* Fungi Col. 2928; *Ohio:* Ell & Ev. N.A.F. 1347 (MASS); *North Dakota:* Brenckle, Fungi Dakotenses 329 (TRTC); *South Dakota:* Brookings, Oct. 1895 (FLAS); *Iowa:* Iowa City, 30 Apr 1933, G.W. Martin (TENN 2681); *Nebraska:* Fungi Col. 3422 (MASS, MICH); *Tennessee:* Flora of Tennessee 6095 (TENN); *Georgia:* Lexington, 19 Mar 1938, G. Thompson (TENN 15690; FLAS, GA 6653, 6654); *Mississippi:* Senatobia, 13 Sep 1929 (TENN 6726).

7. **Linocarpon** Sydow & Sydow, Ann. Mycol. **15** : 210. 1917.

Stromatic tissues prosenchymatous, forming narrow blackened clypeus in epidermal cells of host, surrounding apex of perithecium, hyphae inconspicuous in host tissues; perithecia globose depressed, upright, immersed, beak central, conic, short and broad; peridium narrow at base, thickened and merged with clypeus above. Asci elongate, ±cylindroid. Ascospores hyaline, filiform or broader above and tapering to basal end, usually several septate, guttules numerous, parallel or spirally wound in one or two fascicles in the ascus.

Conidial state not known.

In large-leaved monocotyledonous plants, Palmae, Pandanaceae, in warmer regions.

Type species: *Linocarpon pandani* (Sydow & Sydow) Sydow & Sydow.

Petrak (**1952c**) discussed *Linocarpon* and included in the generic synonymy *Gaeumannomyces* von Arx & Olivier and *Halophiobolus* Linder, pro parte. *Gaeumannomyces* is recognizably a separate genus, with the stromatic tissues not forming a clypeus. Kohlmeyer (**1972**) included the species of *Halophiobolus* in *Lulworthia* in his key to members of the Halosphaeriaceae.

1. **Linocarpon pandani** (Sydow & Sydow) Sydow & Sydow, Ann. Mycol. **15**: 210. 1917. FIG. 49

Linospora pandani Sydow & Sydow, Ann. Mycol. **11**: 60. 1913.

Linospora pandani Rehm, Leaflets of Philippine Bot. **8**: 2954. 1916.

Perithecia 440-600 μm diam, 190-220 μm high, beak short and broad, ±100 μm long, ca. 130 μm wide near base and tapering to apex, clypeus equal in diam to perithecium or up to 1 mm wide, rounded or oblong; peridium 20-25 μm wide at base and lower sides, up to 40 μm wide above, merging with clypeal tissues. Asci 60-100 × 6-8 μm. Ascospores 65-100 × 1.5-3 μm, hyaline, ±filiform with rounded apex and slightly tapered base, several septate, multiguttulate, small pulvinate gelatinous appendage over each end, spirally wound in one fascicle in the ascus.

On *Pandanus* spp., Philippine Islands, Hawaii, Porto Rico.

Material examined: *Philippine Islands:* Fungi Malayana, C.F. Baker 545, 152, 152 suppl. 1; *Hawaii:* near Sacred Falls, 1 Dec 1927, C.L. Shear and N.E. Stevens; Wailau Valley, Molokai, 2 Aug 1928, O. Degener; *Porto Rico:* Porto Rican Fungi, B. Fink no. 495 (all BPI).

Linospora palmetto Ell. & Ev. (J. Mycol. **3**: 45. 1887) develops in spots in living leaves of *Sabal palmetto.* It is a smaller species of *Linocarpon,* **Linocarpon palmetto** (Ell. & Ev.) Barr, comb. nov., with ascospores 36-48 × 1.5-2 μm. Type material was collected in Louisiana: Home Place, 11 Dec 1886, A.B. Langlois (NY). This fungus is quite similar in ascospores to *Linocarpon versisporum* (Ell. & Mart.) Petrak (Sydowia **6**: 388. 1952) which also occurs on *Sabal.* According to Ellis and Everhart, North American Fungi 1961, *L. versisporum* develops in dead leaf tissues; the ascospores are broadest in the upper third and taper to the base. *Linocarpon umbelliferarum* Barr (**1961**) differs in several respects from the generic concept and is transferred to *Plagiosphaera* later in this study.

I-2b. **Gnomoniaceae** subfamily **Stegophoroideae** tribe **Diaporthopsieae** Barr, tribus nov.

Perithecia singulara vel pauciaggregata, stromata instar zonae nigrifactae vel disci stromatici; in ramorum arboribus et frutibus, herbarum caulibus, vel monocotyledonibus, vel filicibus. Genus typicus: *Diaporthopsis* Fabre.

Key to Genera of Tribe Diaporthopsieae

1. Ascospores nonseptate, in dicotyledonous or fern hosts.......................... 2
1. Ascospores septate, in monocotyledonous hosts.................................. 3
  2. Ascospores allantoid (cylindroid and curved); in woody branches..... 8. *Xenotypa*
  2. Ascospores ellipsoid fusoid; in herbaceous stalks or in ferns..... 9. *Diaporthopsis*
3. Ascospores fusoid, one to several septate.......................... 10. *Clypeoporthe*
3. Ascospores narrowly clavoid and tapering to base, several septate.. 11. *Gaeumannomyces*

8. **Xenotypa** Petrak, Sydowia **9**: 499. 1955.

Stromatic tissues prosenchymatous, forming effused black areas over large portions of branch, arising beneath outer periderm, prominent sterile stromatic processes arising from surface, conic, columnar, or

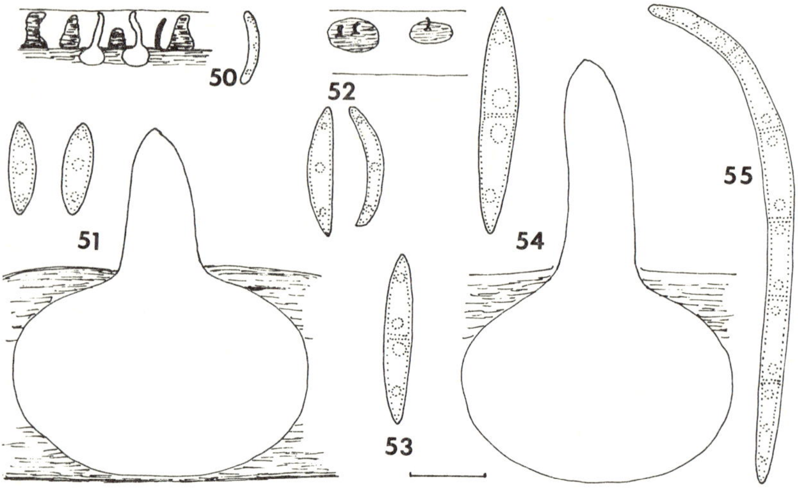

FIGS. 50-55. FIG. 50. *Xenotypa aterrima,* habit of perithecia and sterile stromatic processes, ascospore. FIGS. 51, 52. Species of *Diaporthopsis.* 51. *D. angelicae,* outline of perithecium, ascospores. 52. *D. sepium,* surface view of stroma and perithecial beaks, ascospores. FIGS. 53, 54. Species of *Clypeoporthe.* 53. *C. monocarpa,* ascospore. 54. *C. andropogonis,* outline of perithecium, ascospore. FIG. 55. *Gaeumannomyces graminis,* ascospore. Scale =1 mm, 100 μm, or 10 μm.

irregular, the uppermost thin layer of stroma forming a compact, thin crust, with some distorted periderm cells included, beneath the crust loosely interwoven hyphae among periderm cells; perithecia upright, relatively large, beak elongate, central. Asci numerous, ellipsoid. Ascospores hyaline, allantoid, one celled.

Conidial state not known.

On branches and twigs of deciduous trees.

Type and single species *Xenotypa aterrima* (Fries) Petrak.

Petrak (**1955a**) described this genus to accommodate *Radulum aterrimum* (Fries) Fries, a fungus which was first recognized as an ascomycete rather than a basidiomycete by von Höhnel (**1909a**). The excellent illustrations and notes provided by Pirozynski (**1974**) permitted identification of the two overmature Alaskan specimens cited here, for extension of the range of the fungus to North America. Pirozynski (**1974**) also discussed the nomenclature of the generic name, and proposed that *Xenotypa* be utilized, and *Radulum* Fries be considered a *nomen ambiguum*.

1. **Xenotypa aterrima** (Fries) Petrak, Sydowia **9**: 499. 1955. FIG. 50

Hydnum aterrimum Fries, Syst. Mycol. **1**: 416. 1821.

Radulum aterrimum (Fries) Fries, Elenchus Fung. **1**: 148. 1828.

Eutypa aterrima (Fries) Nannf., Trans. Brit. Mycol. Soc. **20**: 204. 1936.

Racodium turfaceum var. *cornutum* Pers., Mycol. Europ. **1**: 68. 1822.

Eutypa hydnoidea (Fries) v. Höhnel, Sitzungsber. Kaiserl. Akad. Wiss., Math.-Naturwiss. Cl. Abt. 1, **118**: 1465. 1909.

Sphaeronema ? hydnoideum Fries, Svenska K. Akad. Vet. Handl. Ser. 3, **6**: 358. 1818.

Perithecia 330-495 μm diam, depressed globose, ca. 330 μm high, beak 200-440 μm or longer, often bent, relatively broad, tapering to apex. Asci ca. 30 × 6 μm. Ascospores 8-10(-16) × 1.5 μm.

On branches of *Betula* spp., Europe, North America.

Material examined: *Alaska:* W.B. Cooke 42396, 42449 (part in MASS).

9. **Diaporthopsis** H. Fabre, Ann. Sci. Nat. Bot. Ser. 6. **15**: 35. 1883.

Stromatic tissues prosenchymatous, forming blackened marginal zones at surface of substrate and often deep within substrate also, effused and eutypoid in general aspect; perithecia upright, immersed, beaks central, short or elongate, relatively broad. Asci ellipsoid. Ascospores hyaline, fusoid ellipsoid, one celled, usually containing three globules or groups of guttules, appendages short and narrow when formed.

Conidial state: *Phomopsis* (Sacc.) Sacc. spp.

In overwintered stalks of herbaceous plants or rachises of ferns.

Type species: *Diaporthopsis nigrella* (Niessl) H. Fabre =*D. angelicae* (Berk.) Wehm.

Wehmeyer (**1933**) placed *Clypeoporthella* Petrak and *Batschiella* Kirschstein in synonymy under *Diaporthopsis. Batschiella* is a synonym of *Clypeoporthella*, which differs from *Diaporthopsis* in horizontal perithecia and lateral beaks. *Clypeoporthella* is retained as a genus separate from *Diaporthopsis*, as Petrak (**1940**) advocated, and is a member of tribe Hypospilineae. Von Arx and Müller (**1954**) included the European *Aporthytisma* v. Höhnel, with *A. urticae* (Fries) v. Höhnel (Ann. Mycol. **15**: 318. 1917), as a synonym of *Diaporthopsis.* According to the literature (Petrak, **1927**), *Aporhytisma* is indeed closely related to *Diaporthopsis,* but Petrak (**1971a**) retained it as a separate genus.

1. **Diaporthopsis angelicae** (Berk.) Wehm., The Genus *Diaporthe,* p. 228. 1933. FIG. 51

Sphaeria angelicae Berk., Mag. Zool. & Bot. **1837**: 28. 1837.

Sphaeria berkeleyi Desm. in Mont., Ann. Sci. Nat. Bot., ser. 2, **8**: 358. 1837.

Diaporthe berkeleyi (Desm.) Nits., Pyreno. Germ. 273. 1867.

Gnomonia berkeleyi (Desm.) Karst., Mycol. Fenn. **2**: 123. 1873.

Leptosphaeria nigrella Auersw., Mycol. Eur. **5/6**: 12. 1869.

Diaporthe nigrella (Auersw.) Niessl, Beitr. Kennt. Pilze, p. 51. 1872.

Diaporthopsis nigrella (Auersw.) Fabre, Ann. Sci. Nat. Bot., ser. 6, **15**: 35. 1883.

Diaporthe foeniculacea Niessl in Thümen, Contr. Fl. Mycol. Lusit. **2**: 30. 1880.

Diaporthe tetraspora Sacc., Michelia **2**: 250. 1881.

Diaporthe denigrata Winter in Sacc., Syll. Fung. **1**: 649. 1882.

Stromata delimited by blackened marginal zones in epidermal layer and at edge of vascular tissues, effuse in stalks; perithecia separate, scattered, 385-440 μm diam, 250-275 μm high, beaks 165-275 μm or more long, ca. 135 μm wide. Asci 40-50 × 8-10 μm. Ascospores (9-)11-15 (-17) × 3.5-4(-5) μm, fusoid, appendages short and delicate.

Conidial state: stromata containing locules; conidiogenous cells phialidic; conidia 23-31 × 1.5 μm, hyaline, elongate, curved.

In overwintered stalks of species of Umbelliferae, Europe, North America.

Material examined: *North America: British Columbia:* Barr 5584 (MASS).

This collection is the only one seen from North America; Wehmeyer (**1933**) reported the species only from Europe. *Phomatospora angelicae* (Fckl.) Mouton is quite similar to *D. angelicae* and since both occur on similar substrata they may be confused. *Phomatospora angelicae* lacks stromatic tissues, has subepidermal perithecia, short beaks, more nearly cylindroid, stipitate asci which form a hymenium interspersed with narrow paraphyses, and ±uniseriate ascospores, ellipsoid with rounded ends, containing two globules.

Wehmeyer (**1933**) provided a key to species of *Diaporthopsis* and descriptions and illustrations of the species. *Diaporthopsis trinucleata* (Niessl) v. Höhnel (Ann. Mycol. **16**: 114. 1918) on *Eupatorium* and other Compositae, shows blackening above the perithecia but no ventral marginal zones; the ascospores are similar to those of *D. angelicae* with a distinct short appendage on each end (Ontario: Bartholomew, Fungi Col. 2017). *Diaporthopsis apiculosa* (Ell.) Wehm. (The genus *Diaporthe*, p. 231. 1933) on *Erigeron* stalks, has blackened marginal zones delimiting both upper and lower extremes of stromatic tissues, and smaller appendaged ascospores, 7.5-9 × 3-3.5 μm. (New Jersey: Ellis, N.A.F. 787, isotype). *Diaporthopsis pantherina* (Berk.) Wehm. (The genus *Diaporthe*, p. 232. 1933) on *Pteridium aquilinum* in England produces rather isolated small stromatic areas, delimited by blackened marginal zones, and is otherwise similar to *D. angelicae.*

Diaporthopsis therophila (Desm.) v. Höhnel is better placed in *Phomatospora* as *P. therophila* (Desm) Sacc. (Barr, **1961**). *Diaporthopsis appendiculata* Wehm. (The genus *Diaporthe*, p. 233. 1933) on *Aster*, with oblique to horizontal perithecia often forming small groups, with ±lateral beaks, is transferred to *Clypeoporthella* (see that genus).

Mazzantia sepium Sacc. & Penzig does not have the sclerotial stromatic tissues typical of *Mazzantia*, as v. Höhnel (**1918**) had previously observed. This species is here transferred to *Diaporthopsis.*

2. **Diaporthopsis sepium** (Sacc. & Penzig) Barr, comb. nov. FIG. 52

Mazzantia sepium Sacc. & Penzig, Michelia **2**: 609. 1882.

Stromata small, 1-1.5 × 0.5-1 mm or at times two or more fused, delimited by blackened marginal zones, interior of hyaline hyphae mixed with substrate cells; perithecia one or few in a stroma, upright, 250-300 μm diam, 200 μm high, beak central, 200-250 μm long, 100 μm wide, often bent or curved near apex. Asci 45-60 × 9-12 μm. Ascospores 15-22.5 × 3.5-4.5 μm, fusoid, inequilateral or curved, one celled, containing three globules, no appendages formed.

In overwintered stalks of *Convolvulus sepium* L., Europe, North America.

Material examined: *North America: Ontario:* Bartholomew, Fungi Col. 4127 (MASS); *Massachusetts:* Barr 6251 (MASS); *New York:* Hyde Park, 1971, H. E. Ahles (MASS).

The ascospores in the North American collections are larger than described from Europe (10-12 × 2.5-3.5 μm), as Ellis and Everhart (**1892**) earlier noted for a New York collection.

From the description and illustrations of *Mazzantia yukawana* (Hino & Katumoto) Kobayashi (Bull. Gov. Forest Exp. Sta. Tokyo **226**: 155. 1970), this species too belongs in *Diaporthopsis*. The stromatic tissues are prosenchymatous and include substrate cells.

10. **Clypeoporthe** v. Höhnel, Sitzungsber. Kaiserl. Akad. Wiss., Math.-Naturwiss. Cl., Abt. 1, **128**: 584. 1919.

Stromatic tissues prosenchymatous, as interwoven hyphae loosely surrounding perithecia, at times forming epidermal clypeus or blackening substrate surface; perithecia upright, immersed, beaks central, elongate, broad; peridium narrow, covered with loose weft of brown hyphae. Asci oblong or ellipsoid. Ascospores hyaline, narrowly ellipsoid fusoid, septum median.

Conidial state known only for *C. iliau*, as *Melanconium* Link ex Fr.

In stalks and leaves of monocots.

Type species: *Clypeoporthe monocarpa* v. Höhnel.

This genus has had very little notice taken of it. Wehmeyer (**1933**) did not treat *C. monocarpa*, but left the second species of the genus, *C. bambusae* (Pat.) v. Höhnel in *Clypeoporthe*. He noted that the perithecia were solitary beneath a localized darkened clypeus and that ascospores were hyaline becoming light brown, 21.5-25 × 5-6 μm. I accept the genus and extend the concept to include two species which may have blackened substrate surface, but which do not form individual clypei.

1. **Clypeoporthe monocarpa** v. Höhnel, Sitzungsber. Kaiserl. Akad. Wiss., Math.-Naturwiss. Cl., Abt. 1, **128**: 584. 1919. FIG. 53

Stromatic tissues blackening stalks as longitudinal streaks; perithecia separate or few grouped together, 440-700 μm diam, beak 135-200 μm or longer, 220-360 μm wide; peridium thin, beneath epidermal clypeus up to 80 μm wide. Asci 60-70 × 9-12 μm. Ascospores 20-29 × 3-5 μm, straight or curved, biseriate.

In a large "sweet grass" "Sussgras," Java.

Material examined: *Java:* Botanical Garden, Buitenzorg, 1907-8, von Höhnel (FH, type).

2. **Clypeoporthe andropogonis** (Ell. & Ev.) Barr, comb. nov. FIG. 54

Gnomonia andropogonis Ell. & Ev., North Amer. Pyreno. p. 324. 1892.

Perithecia 200-500 μm diam, 150-250 μm high, thickly grouped or scattered, beaks 150-900 μm long, 80-150(-230) μm wide near base, at times encrusted with bright yellowish crystalline material; peridium surrounded by loosely interwoven hyphae. Asci 44-55 × 6.5-11 μm. Ascospores 22-35 × 3.5-5.5 μm.

In culms of *Andropogon scoparius* Michx., *A. virginicus* L., North America.

Material examined: *New Jersey:* Newfield, 25 Dec 1890, J.B. Ellis, 4 Jan 1891, 16 Oct 1892 (all marked paratype, NY); 1857(?), 4 Jun 1893, 7 Feb 1897 (NY); *New York:* Loudonville, Albany Co., 7 Aug 1944, H.D. House (NY, NYS); E. of Saratoga Lake, 9 Sep 1944, H.D. House (NYS); *Kansas:* Agronomy Farm, Manhattan, Riley Co., 23 Oct 1950, C.T. Rogerson (BPI); 1 Oct 1951 (NY); *Mississippi:* Starkville, 23 Mar 1895, S.M. Tracy (BPI); 10 Oct 1894 (BPI, NY).

Rogerson (**1953**) suggested that this fungus may be the cause of a crown rot disease of *Andropogon*.

3. **Clypeoporthe iliau** (Lyon) Barr, comb. nov.

Gnomoniia iliau Lyon, Hawaiian Sugar Planter's Assoc. Path. & Phys. Ser. Bull. **11**: 32. 1912.

Perithecia 325-495 μm diam, 240-385 μm high, beaks 275-330 (-550) μm long, 130-137 μm wide near base; peridium surrounded by loosely interwoven hyphae. Asci 50-60(-80) × 8-14 μm. Ascospores 14-20 (-30) × 3-5(-7) μm.

In culms of *Saccharum officinarum* L., North America.

Material examined: *Louisiana:* Port Allen, 14 Sep 1912, C.W. Edgerton (TRTC, as Sydow, Fungi exotici exs. 512).

Edgerton (**1913**) described and illustrated this species from Louisiana specimens. It causes a stem rot of sugar cane. The conidial state, *Melanconium iliau* Lyon, has stromata 500-700 μm, containing a single locule or chambered structure, conidia 15-28 × 7-10 μm, ellipsoid to ovoid, dark brown, one celled, with granular contents, borne on hyaline conidiophores.

11. **Gaeumannomyces** von Arx & Olivier, Trans. Brit. Mycol. Soc. **35**: 32. 1952.

Stromatic tissues slightly developed, prosenchymatous, forming a mycelial mat on stolons and leaf sheaths and plate-like layer between culm and inner side of sheath; hyphae bearing simple or lobed hyphopodia; perithecia separate, immersed, often oblique, beaks oblique or ±central, short and broad; peridium relatively wide. Asci elongate, ±cylindroid. Ascospores hyaline or light yellowish, ±filiform or narrowly clavoid elongate, tapering to base, several septate, not constricted, straight or curved, contents guttulate or granular.

Conidial state: *Phialophora*-like (in *Phialophora radicicola* Cain complex): produced in culture on hyphae and from germinating ascospores; conidia phialidic, 4-8 × 1.5 μm, hyaline, falcate.

On roots, culms, and leaf sheaths of genera of Gramineae, Cyperaceae.

Type species: *Gaeumannomyces graminis* (Sacc.) von Arx & Olivier.

Gaeumannomyces graminis has been known for a long time as *Ophiobolus graminis.* Fitzpatrick, Thomas, and Kirby (**1922**) described the fungus causing take-all of wheat and concluded that *Sphaeria cariceti* Berk. & Broome was the earliest name for the species. Walker (**1972**) studied a number of fungi related to *G. graminis* and retained the epithet *graminis* for the cereal take-all fungus. Jones (**1926**) provided a detailed description of perithecial development in *Ophiobolus graminis.* He noted that many similarities existed between this species and species of *Gnomonia* and concluded that the species could not be assigned to the Pleosporaceae but that it was more closely allied to the Gnomoniaceae. Other authors also questioned the position of this species in *Ophiobolus.*

Von Arx and Olivier (**1952**) provided the generic name *Gaeumannomyces* and compared *G. graminis* with other scolecosporous genera in the Diaporthaceae. Petrak (**1952c**) reduced *Gaeumannomyces* to synonymy with *Linocarpon,* and Munk (**1957**) and Kobayashi (**1970**) concurred. However, Dennis (**1960**) and Walker (**1972**) disagreed and recognized *Gaeumannomyces* as a separate genus. Walker (**1972**) gave detailed descriptions and illustrations of the varieties of *G. graminis,* and reported the presence of brown, simple or lobed hyphopodia on the superficial runner hyphae. Weste (**1972**) studied ascospore germination and found that either germ tubes and hyphae or microconidia may be produced. Conidia are also formed from short phialides on hyphae in culture or on roots.

Linocarpon differs from *Gaeumannomyces* particularly in the small clypeus which surrounds the beak; perithecia are upright and beaks are central consistently. *Linocarpon* is placed in the tribe Stegophoreae while *Gaeumannomyces* seems most closely related to genera in the tribe Diaporthopsieae. The perithecia in *Gaeumannomyces* species are often oblique, and the peridium is relatively thick. Delicate sterile tissues are frequently found in the centrum, broad at the base and tapering to the tip, septate into elongate cells. These structures are clearly illustrated in Shoemaker's (**1974**) contribution # 37 to Fungi Canadenses. Jones (**1926**) observed that these 'paraphyses' were formed by separation and elongation of pseudoparenchymatous centrum cells, in rows parallel to the long axis of the perithecium.

1. **Gaeumannomyces graminis** (Sacc.) v. Arx & Olivier, Trans. Brit. Mycol. Soc. **35**: 32. 1952. FIG. 55

Rhaphidophora graminis Sacc., Nuov. Giorn. Bot. Ital. **7**: 307. 1875.

Ophiobolus graminis (Sacc.) Sacc., Syll. Fung. **2**: 349. 1883.

Ophiochaeta graminis (Sacc.) Hara, J. Plant Prot. **3**: 342. 1916.

Perithecia globose or ovoid, 200-500 μm diam or 200-300 × 150-200 μm; beak short and broad, 100-250 μm long, 80-180 μm wide; superficial hyphae brown, 2.5-6 μm wide, septate, bearing hyphopodia, 17-27 μm diam or up to 30 × 20 μm. Asci 70-110 × 10-15 μm. Ascospores 60-95(-141) × 2-5 μm, up to 13 septate, parallel in a fascicle in the ascus.

Gaeumannomyces graminis var. **graminis**: Causing rice brown sheath rot; hypophodia lobed or simple; ascospores 70-115 × 2-4.5 μm, three to five septate.

Material examined: On *Oryza sativa* L. *Arkansas:* Stuttgart, Fall 1931, E.C. Tullis (BPI, as *Leptosphaeria salvinii* Catt.).

Gaeumannomyces graminis var. **avenae** (Turner) Dennis, British Cup Fungi, p. 202. 1960: Causing oat take-all; hyphopodia simple; ascospores 65-141 × 3-5 μm, up to 13 septate.

Gaeumannomyces graminis var. **tricici** Walker, Trans. Brit. Mycol. Soc. **58**: 439. 1972: Causing wheat take-all; hyphopodia simple; ascospores 60-118 × 3-5 μm, up to nine septate.

Several closely related species are known in the genus on members of the Cyperaceae.

I-3. **Gnomoniaceae** subfamily **Mamianioideae** Barr, subfam. nov.

Perithecia singulara vel aggregata, stromata contexta pseudoparenchymata. Genus typicus: *Mamiania* Cesati & de Notaris (two tribes).

I-3a. Tribe **Mamianieae**

Perithecia single or few in small stromata; in leaves, petioles, and branches of deciduous trees and shrubs and in herbaceous stalks.

Key to Genera of Tribe Mamianieae

1. Stroma composed of thick-walled sclerotial cells; ascospores nonseptate, allantoid or ellipsoid .... 12. *Mazzantia*

1. Stroma composed of thin-walled, nonsclerotial cells; ascospores not allantoid....... 2

  2. Ascospores nonseptate, ellipsoid or ovoid; stroma small, containing one or few perithecia.. 13. *Mamianiella*

  2. Ascospores septate 3

3. Ascospores apiosporous, ellipsoid or ovoid; stroma relatively large, containing a number of perithecia.. 14. *Mamiania*

3. Ascospore septum median, ellipsoid fusoid; stroma small, containing one or few perithecia.. 15. *Ditopellopsis*

The extralimital genera *Phylloporthe, Lambro,* and *Ditopellina* may be included in this tribe.

Phylloporthe H. Sydow (Ann. Mycol. **23**: 348. 1925) with type species *P. vernoniae* Sydow, is parasitic in living leaves of *Vernonia* spp. in tropical America. The pseudoparenchymatous stromata are erumpent from the leaf. Müller and von Arx (**1962**) described and illustrated this species. They also included *Uleoporthe* Petrak as a synonym of *Phylloporthe*; however, the stromatic tissues in *Uleoporthe* are prosenchymatous and that genus is considered to belong in subfamily Stegophoroideae.

Lambro Raciborski (Parasit. Algen u. Pilze Javas **2**: 13. 1900) is similar to *Mamiania* in having pseudoparenchymatous stroma in leaves and in possessing apiospores. It is a parasitic fungus which forms necrotic spots in leaves of *Sterculia* sp. in Indonesia, and the stroma is more fleshy and lighter colored than in *Mamiania.* Müller and von Arx (**1962**) described and illustrated *L. insignis* Racib. They also united the

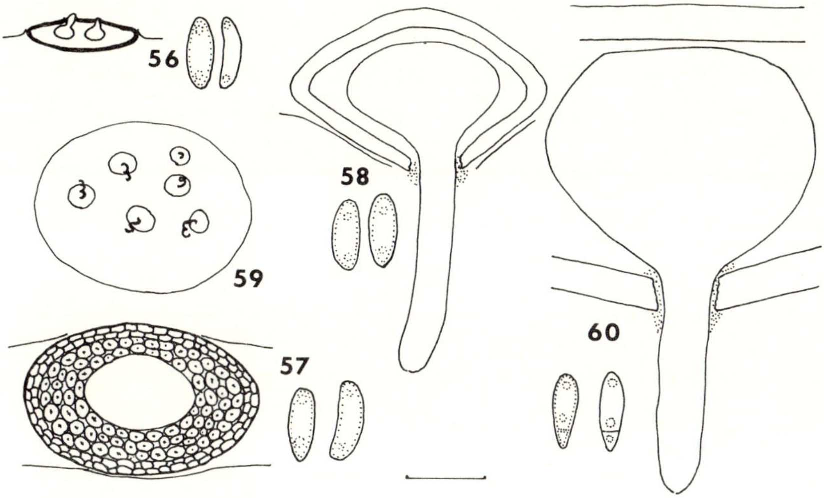

FIGS. 56-60. FIGS. 56, 57. Species of *Mazzantia*. 56. *M. galii,* habit of perithecia in stroma, ascospores. 57. *M. biennis,* details of uniperithecial stroma, ascospores. FIGS. 58, 59. Species of *Mamianiella*. 58. *M. coryli* var. *coryli,* outline of uniperithecial stroma, ascospores. 59. *M. coryli* var. *spiralis,* surface view of coiled perithecial beaks. FIG. 60. *Mamiania fimbriata,* outline of perithecium and portion of stroma, ascospores. Scale =1 mm, 100 μm, or 10 μm.

genera *Stegophora* and *Lambro* under the latter, earlier name. *Stegophora* is retained for *S. ulmea* in my concept of generic segregations.

Ditopellina Reid & Booth (Canad. J. Bot. **45**: 1481. 1967) has as the type species *D. saccardiana* (Trav. & Spessa) Reid & Booth on *Cinnamomum* sp., Portugal. The pseudoparenchymatous stroma is expanded beneath and at the sides of one or few upright perithecia. Beaks are central and short conic. The elongate fusoid ascospores are distinctive in the Mamianieae.

12. **Mazzantia** Mont., Syll. Gen. p. 215. 1856.

Paramazzantia Petrak, Ann. Mycol. **25**: 232. 1927.

Clypeocarpus Kirschst. Hedwigia **80**: 126. 1941.

Stromatic tissues pseudoparenchymatous, sclerotial, stroma externally of two to four layers of blackened cells, white within, inner stromatic cells thick walled with small lumen; perithecia one or few in stroma, upright, beak central, as minute papilla or short to elongate. Asci ellipsoid. Ascospores hyaline, ellipsoid or allantoid, straight or one side flattened, one celled, a globule near each end.

Conidial state: *Mazzantiella* v. Höhnel or *Paramazzantiella* Petrak: locule developing in stroma similar to perithecial stroma; conidia hyaline, elongate or cylindroid, one celled.

In herbaceous stalks and leaves.

Type species: *Mazzantia galii* (Fries) Mont.

Paramazzantia Petrak was erected for *Laestadia biennis* Dearness. This species, in overwintered leaves of *Solidago juncea*, has sclerotial stromata as in *M. galii*, and differs from it only in the extremely short beak. Separation into two genera of these similar and distinctive fungi is not practicable, thus *Paramazzantia* is placed in synonymy with *Mazzantia*.

Von Arx and Müller (**1954**) synonymized *Paramazzantia* with *Heteropera* Theissen (Ann. Mycol. **14**: 423. 1916). Theissen (**1916**) described the type species, *H. borealis* (Sacc.) Theissen, as a member of the Coronophoraceae. This species was based upon *Physalospora borealis* Sacc. Theissen provided a detailed description of the fungus, and noted that superficially it appeared as fruiting bodies of *Pyrenophora* - *Scleroplea*, large, matt-black, not ostiolate. The thick wall surrounded a thin-walled perithecium which produced a small beak to the surface. Von Arx and Müller (**1954**) interpreted the thick wall as a sclerotial stroma and accepted *Heteropera* as a member of the Diaporthaceae. However, Petrak (**1960**) recognized that *H. borealis* was based upon a fungus identical with *Cryptonectriopsis biparasitica* (v. Höhnel) v. Höhnel which developed within empty ascocarps of a pleosporaceous fungus. Petrak later (**1971a**) reiterated that *Paramazzantia* and *H. borealis* were entirely different entities.

The type species of *Cryptonectriopsis* (v. Höhnel) v. Höhnel (Ann. Mycol. **17**: 131. 1919), included in the synonymy of *Heteropera* by von Arx and Müller, was originally described as *Hyponectria* (subg. *Cryptonectriopsis*) *biparasitica* von Höhnel (Ann. Mycol. **16**: 36. 1918) and was based upon Rehm's Ascomycetes n. 1523, determined as *Phomatospora ovalis* (Pass.) Sacc. Passerini's original collection, as v. Höhnel recognized, was a pycnidial fungus parasitic in *Heteropatella lacera*. Von Höhnel erected *Mycosticta* with *M. ovalis* (Pass.) v. Höhnel (Ann. Mycol. **16**: 36. 1918) for this fungus. Rehm's n. 1523 differed from Passerini's fungus and v. Höhnel described it as a parasite in old ascocarps of *Leptosphaeria dolioloides*. From the description it appears to be a species of *Phomatospora*, probably not distinct from *H. borealis*. *Cryptonectriopsis* then is synonymous with *Phomatospora*, and is not related to *Mazzantia*.

Clypeocarpus alpinus Kirschst., the type species of the genus, according to von Arx and Müller (**1954**) is identical with *Mazzantia napellii* (Fries) Sacc. Von Höhnel (**1918**) observed that while both *M. galii* and *M. napellii* possessed a sclerotial stroma, *M. sepium* Sacc. & Penzig did not. This latter species has been transferred to *Diaporthopsis* earlier in this paper.

1. **Mazzantia galii** (Fries) Mont., Syll. Crypt. p. 246. 1856. FIG. 56

Sphaeria galii Fries, Elenchus Fung. **2**: 105. 1828.

Sphaeria aparines Cast., Catal. plantes Marseille, p. 171. 1845.

Sphaeria molluginis Otth, Mitth. Naturf. Ges. Bern, p. 57. 1868.

Physalospora molluginis (Otth) Sacc., Hedwigia **35**: 27. 1896.

Stromata 1-2 mm long, 0.5-1 mm wide, 220-275 μm deep, subepidermal, finally erumpent, base attached to cells of cortex; perithecia 180-330 μm diam, (90-)115-220 μm high, beaks short and broad, ca. 130 μm wide and long. Asci 25-35 × 6-9 μm. Ascospores 9-11 × 3.5-4.5 μm.

In overwintered stalks of *Galium* spp., Europe, North America.

Material examined: *Europe:* Ex Flora moravica, Mähr-Weisskirchen, 1912, F. Petrak; Fungi Eichleriani, F. Petrak, n. 221; Flora Bohemiae et Moraviae exs. n. 109; Mycotheca generalis 266; Thümen, Mycotheca univ. 1952, 71; Herbier Barbey-Boissier 720, 767 (all BPI). *North America: Wisconsin:* Madison, 21 Aug 1952, H.C. Greene (BPI).

2. **Mazzantia biennis** (Dearness) Barr, comb. nov. FIG. 57

Laestadia biennis Dearness, Mycologia **8**: 98. 1916.

Paramazzantia biennis (Dearn.) Petrak, Ann. Mycol. **25**: 233. 1927.

Guignardia biennis (Dearn.) Sacc., Syll. Fung. **24**: 780. 1928.

Heteropera biennis (Dearn.) von Arx & Müller, Beitr. Kryptogamenfl. Schweiz **11**(1): 361. 1954.

Stromata 250-360 μm diam, 150-250 μm high, perithecia single in stroma, 140-250 μm diam, 100-200 μm high. Asci 40-55 × 6-11 μm. Ascospores 7.5-11 × 3-4.5 μm. Conidia 24-30 × 1 μm.

In overwintered leaves of *Solidago juncea* Ait., North America.

Material examined: *Ontario:* Bartholomew, Fungi Col. 5030 (MASS, MICH); East Springbank, 17 Aug 1904, J. Dearness; Springbank, Mar 1911, 13 May 1911, 15 Apr 1911, 20 May 1911, 19 May 1912, 18 May 1912 (lectotype) (all in Dearness Herb. in DAOM).

3. **Mamianiella** v. Höhnel, Ann. Mycol. **16**: 102. 1918.

Stromatic tissues pseudoparenchymatous, blackened externally, whitish or light brown internally, stromata small, perithecium single or two or three fused together and stroma enlarged, occasionally containing up to nine perithecia; beaks erumpent hypophyllously, elongate, straight, curved, or spirally twisted, surrounded at base by collar of whitish material. Asci ellipsoid, floating free. Ascospores hyaline, one celled, ovoid, often inequilateral, minute guttule at each end, wall smooth, overlapping biseriate in the ascus.

Parasitic, usually numerous in small yellowish or brown spots, maturing in living leaves of *Corylus* spp., widespread in Europe and North America.

Type and single species: *Mamianiella coryli* (Batsch ex Fries) v. Höhnel.

Mamianiella coryli was segregated from *Mamiania fimbriata* because of the one-celled ascospores of the former versus apiospores of the latter species. Von Arx and Müller (**1954**), as well as others, suggested retaining both species in one genus. The emphasis on ascospore septation in this study calls for the retention of two genera. *Mamianiella* tends to have more numerous small stromata containing one or few perithecia in a leaf spot while in *Mamiania*, one or few larger stromata containing numerous perithecia are formed in a leaf spot.

1. **Mamianiella coryli** (Batsch ex Fries) v. Höhnel, Ann. Mycol. **16**: 102. 1918. var. **coryli** FIG. 58

Sphaeria coryli Batsch ex Fries, Syst. Mycol. **2**: 426. 1823.

Mamiania coryli (Batsch ex Fries) Ces. & de Not., Comment. Soc. Critt. Ital. **1**(4): 211. 1862.

Gnomonia coryli (Batsch ex Fries) Auersw., Mycol. europ. **5/6**: 22. 1869; Fuckel, Symb. Mycol. p. 120. 1870.

Gnomoniella coryli (Batsch ex Fries) Sacc., Syll. Fung. **1**: 419. 1882.

Stromata 300-330 μm diam, or up to several mm diam when several fused together, 200-300 μm high, perithecia 280-380 μm diam, beaks 245-420 μm long, 48-70 μm wide near base, scarcely tapering to blunt apex, straight or slightly curved toward apex. Asci 33-50 × 7-9 μm. Ascospores 6.5-11 × 3-4.5 μm.

In living leaves of *Corylus avellana* L., Europe; *C. americana* Walt., *C. cornuta* Marsh., *C. californica* (A. DC.) Rose, mid- and western North America, Newfoundland.

Material examined (representative collections): *Europe:* Herb. Barbey-Boissier 81 (TRTC); Eriksson, Fungi parasitici Scand. 291 (GA); Flora Exs. Austro-Hungarica 783 (DAOM); Mougeot & Nestler, Stirpes Cryptogamae Vogeso-Rhenanae 877 (DAOM). *North America: Newfoundland:* Bay of Islands, Oct 1897, A.C. Waghorne (DAOM 36350); Newfoundland Fungi 236 (DAOM); *Manitoba:* Brandon, 15 Sep 1917, W.P. Fraser (DAOM 36342); *Saskatchewan*: Moosomin, 8 Aug 1929, B.J. Sallans & I.L. Conners (DAOM 592); *British Columbia:* Canadian Fungi 99 (DAOM); Quesnel, 15 Sep 1956, W.G. Ziller (DAOM); *Minnesota:* Brainerd, Aug 1890, C. Macmillan & E.B. Sheldon (DAOM 36338); *Montana:* Canyon Creek, 12 Oct 1909, G.G. Hedgcock (BPI); *Washington:* Bartholomew, Fungi Col. 4422 (FLAS, MASS, NY); Ellis, N.A.F. 598b (MASS); *Oregon:* Flora of Oregon 1700 (NY, TRTC), 1710 (TRTC), 2641 (TRTC), 2648 (TRTC); *Colorado:* Bartholomew, Fungi Col. 3237 (FLAS, MASS, NY); *Utah:* La Motte Peak, M.E. Jones (NY); *California:* California Fungi 312 (TRTC).

Mamianiella coryli as described on *Corylus avellana* from Europe has beaks which are straight or curved toward the apex. In North America this taxon occurs from the midwestern and western provinces of Canada and states of United States, and also from Newfoundland. Peck's variety *spiralis*, similar in microscopic characters to the typical variety, has beaks which are spirally coiled, usually twice. The variety was described on *Corylus cornuta* from New York and occurs on this species and *C. americana* throughout northeastern North America and westward as far as Alberta. Both varieties are found in the midwestern region of the continent.

2. **Mamianiella coryli** (Batsch ex Fries) v. Höhnel var. **spiralis** (Peck) Barr, comb. nov. FIG. 59

Sphaeria coryli var. *spiralis* Peck, New York State Museum Rept. **34**: 57. 1881.

Differing from var. *coryli* in the spirally twisted beaks.

In living leaves of *Corylus cornuta* Marsh, *C. americana* Walt., eastern to midwestern North America. Also reported on *Betula* sp. (Quebec: Perce, 12 Aug 1929, G.P. Clinton, BPI) and *Alnus* sp. (North Carolina: Mt. Pisgah, 6 Jun 1935, L.R. Hesler, BPI).

Material examined (representative collections): *North America: Nova Scotia:* Nine Mile River, 27 Jul 1932, K.A. Harrison (DAOM 1888); *New Brunswick:* Fredericton, 17 Aug 1953, A.G. Davidson (DAOM 40188); *Quebec:* Cain 11290 (TRTC); *Ontario:* Ell. & Ev. Fungi Col. 117 (FLAS, NY); Canadian Fungi 227 (DAOM); *Manitoba:* Victoria Beach, Aug. 1956, J. Morgan-Jones (TRTC); *Saskatchewan:* Madge Lake, 8 Aug 1939, R.C. Russell (DAOM 5791); *Alberta:* Battle, 28 Aug 1927, A.H. Brinkman (DAOM 4092); *Maine:* Rowe Pond, Sep 1915, J.F. Collins & P. Spaulding (BPI); *New Hampshire:* Ellis N.A.F. 598; *Massachusetts:* Seymour & Earle Economic Fungi 174 (MASS); *Connecticut:* Easton, July 1892, L.N. Johnson (BPI); *New York:* Osceola, Aug. C.H. Peck (type of *Sphaeria coryli* var. *spiralis*, NYS); *Michigan:* La Sage Lake, Isle Royale, 2 Aug 1930, C.A. Brown (TRTC); *Illinois:* English Prairie, 20 Aug 1881, A.B. Seymour (BPI); *Wisconsin:* Albany, 10 Jul 1955, H.C. Greene 1833 (DAOM).

Gnomoniella coryli var. *circinata* Dearness & Bisby in Bisby, Buller & Dearness, The Fungi of Manitoba, p. 76. 1929, was described as differing in having circinate perithecia in smaller spots and more strongly developed stroma. The variety does not appear to be more than a form of *M. coryli* var. *coryli*.

14. **Mamiania** Ces. & de Not., Comment. Soc. Critt. Ital. **1**(4): 210. 1863.

Stromatic tissues pseudoparenchymatous, blackened at surface, whitish internally; perithecia few or numerous (up to 30), rarely solitary in a stroma, depressed globose, upright, beak central, erumpent hypophyllously, black, short to elongate, straight or curved, often surrounded at base by white collar. Asci ellipsoid. Ascospores hyaline, ellipsoid, or nearly ovoid, straight or inequilateral, septum near base (apiosporous), contents with one or two globules, overlapping biseriate in the ascus.

Developing in yellowish or brown indeterminate spots in living leaves, maturing in living leaves or after overwintering, of *Carpinus* spp., Europe, Asia, North America.

Lectotype: *Mamiania fimbriata* (Pers. ex Fries) Ces. & de Not.

Cesati and de Notaris (**1863**) noted that their generic diagnosis was based on *Sphaeria fimbriata* Pers. ex Fries, *S. strumellaeformis* de Not., and *S. hystrix* Tode ex de Not., and that *S. coryli* Batsch ex Fries was provisionally inserted into the genus. These four and a fifth species, *M. spiculosa* (Fries) Ces. & de Not. were transferred; several other species were also listed as possibly related. *Mamiania fimbriata* is the accepted lectotype species; *S. hystrix* is *Cryptodiaporthe hystrix* (Tode ex Fries) Petrak, and *S. strumellaeformis* is a synonym of *Diaporthe strumella* (Fries) Fuckel. *Sphaeria coryli* is the type species of *Mamianiella* with one-celled ascospores. *Mamiania spiculosa* (Fr.) Ces. & de Not. is a synonym of *Diaporthe fasciculata* Nits. according to Winter (**1887**), but is *Diaporthe spiculosa* (Alb. & Schw. ex Fries) Nits., according to Wehmeyer (**1933**).

The name of *Kubinya* Schulz. appears as a synonym of *Gnomonia* (*Mamiania*) in Saccardo's Sylloge Fungorum, Index Univ. to. vol. **16**: 1263. 1902. The name is mentioned by Lindau in Engler and Prantl (**1897**), p. 448 as a synonym of *Mamiania* (written as *Kubingia* Schulzer). I have no further information about this name.

Von Arx (**1951**) noted that *Mamiania* differed from "typical" Gnomoniaceae by the type of stromatic tissues, and suggested that it belonged with *Mazzantia* and related forms in a separate family.

1. **Mamiania fimbriata** (Pers. ex Fries) Ces. & de Not., Comment Soc. Critt. Ital. **1**(4): 211. 1863. FIG. 60

Sphaeria (Confertae) *fimbriata* (Pers. ex Fries, Syst. Mycol. **2**: 436. 1823.

Gnomonia fimbriata (Pers. ex Fries) Fuckel, Symb. Mycol. p. 119. 1870.

Gnomoniella fimbriata (Pers. ex Fries) Sacc., Syll. Fung. **1**: 419. 1882.

Sphaeria spiculosa Batsch ex Fries, Syst. Mycol. **2**: 436. 1823.

Mamiania spiculosa (Batsch) Trav., Flora Crypt. Ital. **2**: 167. 1905.

Sphaeria carpini Hoffm. ex Fries, Syst. Mycol. **2**: 436. 1823.

Stromata 550 μm diam, or 1.5-3 mm diam and 500-600 μm deep; perithecia 350-400 μ diam, 300-350 μm high, beak 300-500 μm long, 60-80 μm wide near base; peridium 20-28 μm wide, light brown. Asci 42-55 × 6-9 μm. Ascospores (7-)9-11 × 3-4 μm.

Common although inconspicuous in living leaves of species of *Carpinus* in Europe, Asia, North America; occasionally in leaves in *Ostrya virginiana* (Mill.) K. Koch (New York: Alcove, Sep 1893, C.L. Shear (BPI, TENN, TRTC).

Material examined (representative collections): *Europe:* Herbier Barbey-Boissier 199; Herbier Müller Arg. 1896; Herb. Mycol.Romanicum; Flora Danica; *North America: Ontario:* Ell. & Ev. N.A.F. 2361; Dearness Canadian Fungi 230; *Vermont:* Ell. & Ev. Fungi Col. 935; *New York:* Shear's New York Fungi 333.

In *Mamiania alni* Miller & Thompson, the ascospores have a median septum. This species is transferred to *Ditopellopsis.*

15. **Ditopellopsis** Reid & Booth, Canad. J. Bot. **45**: 1479. 1967.

Stromatic tissues pseudoparenchymatous, stroma surrounding single or few (two to five) perithecia; perithecia upright, beaks centrally erumpent, short or elongate, relatively broad, sometimes fused when more than one perithecium within a stroma. Asci ellipsoid, floating free. Ascospores hyaline, narrowly ellipsoid fusoid, septum median, usually two globules per cell, appendages short and blunt or elongate, overlapping biseriate in the ascus.

In overwintered branches or leaves of hardwood shrubs or trees or large herbaceous stalks, North America.

Type species: *Ditopellopsis clethrae* Reid & Booth.

This genus was erected for a species in which the pseudoparenchymatous stroma closely surrounded the upright perithecium, the beaks were short and broad, and the ascospores had a median septum. Two other species transferred to the genus extend the concept somewhat: *D. racemula* occasionally has more than one perithecium within a stroma and *D. alni* has a conspicuously elongate beak. *Ditopellopsis sophorae* Kobayashi (Bull. Govern. Forest Exp. Station **226**: 48. 1970) according to the description and illustrations is a fungus belonging to the Melanconidaceae.

1. **Ditopellopsis clethrae** Reid & Booth, Canad. J. Bot. **45**: 1479. 1967.
FIG. 61

Stroma 550-715 μm diam, 275-355 μm high; perithecia up to 500 μm diam, single in stroma, beaks 104-120 μm wide, 175-250 μm long. Asci 21-23 × 7.5-10 μm (30-40 μm long according to Reid and Booth), apical annulus narrow. Ascospores (9-)10-14(-16) × 2.5-3.5(-4) μm, usually curved, appendages short, delicate, chitinoid.

In dead branch of *Clethra alnifolia* L., Maryland.

Material examined: *Maryland:* Beltsville, 6 Jun 1936, C.L. Shear (BPI, holotype).

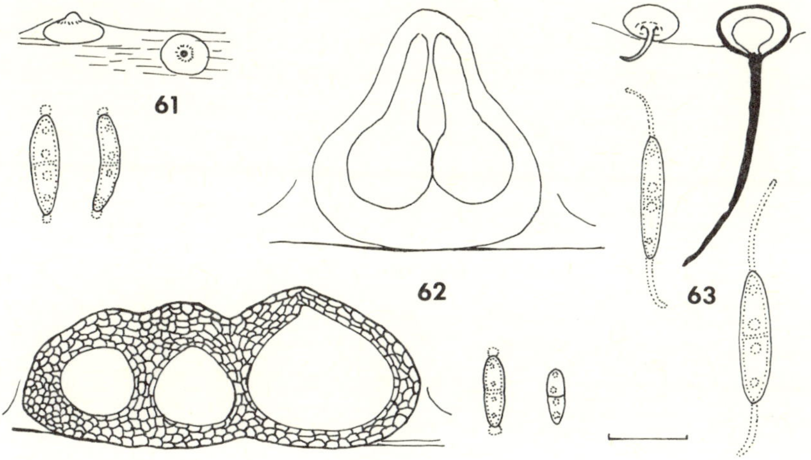

FIGS. 61-63. Species of *Ditopellopsis*. 61. *D. clethrae*, habit of uniperithecial stromata, ascospores. 62. *D. racemula*, outline of perithecia in stroma, details of stroma, ascospores. 63. *D. alni*, habit of uniperithecial stromata, ascospores. Scale =1 mm, 100 μm, or 10 μm.

2. **Ditopellopsis racemula** (Cooke & Peck) Barr, comb. nov. FIG. 62

Sphaeria (Caulicolae) *racemula* Cooke & Peck in Peck, New York State Museum Rep. **26**: 87. (for 1872) 1874 (*nomen nudum*); **29**: 65. (for 1875) 1878.

Diaporthe racemula (Cooke & Peck) Sacc., Syll. Fung. **1**: 691. 1882.

Stromata up to 500 μm diam, immersed, pseudoparenchymatous, blackened externally; perithecia 140-250(-300) μm diam, 120-200 μm high, one or few in a stroma, beaks up to 130-200 μm long, 70-80 μm diam near base and tapering slightly to rounded apex; peridium ca. 12 μm wide, of compressed pallid cells. Asci 24-38.5 × 5.5-8 μm, apical annulus shallow. Ascospores 7.5-11 × 2-3.5 μm, straight or inequilateral, appendages delicate, hyaline, pulvinate, 1-1.5 μm wide and long.

In overwintered stalks of *Epilobium angustifolium* L., North America.

Material examined: *British Columbia:* Barr 5580b; *Maine:* 3759, 5837b (all MASS); *New York:* Lower Ausable, Adirondack Mts., Jul 1873, C.H. Peck (type); Aiden Lair, Jul, C.H. Peck (both NYS).

Ditopellopsis racemula is closely related to *D. clethrae* and differs from that species mainly by occurrence of few smaller perithecia in a stroma and by slightly smaller ascospores. Wehmeyer (**1933**), who also found *D. racemula* in Nova Scotia, suggested that it might represent a new generic type, and excluded the species from *Diaporthe*.

3. **Ditopellopsis alni** (Thompson & Miller) Barr, comb. nov. FIG. 63

Mamiania alni Thompson & Miller, Mycologia **32**: 9. 1940.

Stromata 500-1000 μm diam, ca. 500 μm high, immersed in leaf tissue, blackened externally, yellowish internally, composed of thick-walled pseudoparenchymatous cells; perithecia single in stroma, depressed globose, 285-400 μm diam, 200-300 μm high, beaks erumpent hypophyllously, often bent, 1-4(-7) mm long, 80 μm or more wide at base and tapering to apex; peridium 16.5 μm wide, yellowish layers of cells. Asci 50-65 × 7-11(-14) μm, apical annulus 4-4.5 μm wide and high. Ascospores 16-18.5(-23) × 3-4 μm, straight or inequilateral, appendages elongate, strap-like, 7.5-14.5 μm long, 1-1.5 μm wide, straight or curved.

In overwintered leaves of *Alnus rugosa* (Du Roi) Spreng., North America.

Material examined: *Georgia:* Agricultural Campus, Athens, 5 Jul 1939, G.E. Thompson & J.H. Miller (type, BPI, GA 6718), also GA 6716, 6717.

Ditopellopsis alni differs from the other two species in foliicolous habit, greatly elongated beaks, larger asci and ascospores, conspicuous apical annulus of the ascus, and elongate appendages on the ascospores.

I-3b. **Gnomoniaceae** subfamily **Mamianioideae** tribe **Endothieae** Barr, tribus nov.

Perithecia aggregata in stromate erumpenti; in ramorum arboribus et frutibus. Genus typicus: *Endothia* Fries.

Key to Genera of Endothieae

1. Ascospores nonseptate, allantoid 2
1. Ascospores septate, ellipsoid or fusoid; stroma composed of thin-walled cells, dark colored externally, dull brownish internally 3
    2. Stroma composed of thick-walled, sclerotial cells, dark colored externally, light colored internally.. 17. *Valseutypella*
    2. Stroma composed of thin-walled cells, or compactly interwoven hyphae, brightly colored externally and internally .... 16. *Endothia*
3. Ascospores elongate fusoid, one to three septate .... 20. *Sillia*
3. Ascospores short ellipsoid or fusoid, one septate 4
    4. Ascospores apiosporous.................................... 18. *Anisogramma*
    4. Ascospore septum median.................................. 19. *Diaporthella*

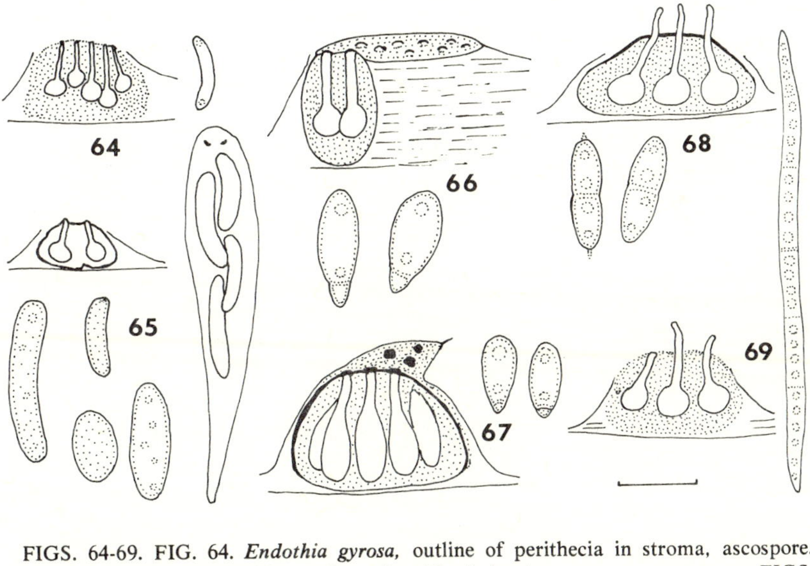

FIGS. 64-69. FIG. 64. *Endothia gyrosa,* outline of perithecia in stroma, ascospore. FIG. 65. *Valseutypella tristicha,* outline of perithecia in stroma, ascus, ascospores. FIGS. 66, 67. Species of *Anisogramma.* 66. *A. virgultorum,* outline of perithecia in stroma, ascospores. 67. *A. anomala,* outline of perithecia in stroma, ascospores. FIG. 68. *Diaporthella aristata,* outline of perithecia in stroma, ascospores. FIG. 69. *Sillia ferruginea,* outline of perithecia in stroma, ascospores. Scale =1 mm or 10 μm.

16. **Endothia** Fries, Summa Veget. Scand. p. 385. 1849.

Stromatic tissues pseudoparenchymatous, stroma diatrypoid, erumpent, brightly colored; perithecia upright, numerous in one layer or several layers, beaks ±central, elongate, erumpent separately to or beyond stroma surface. Asci ellipsoid. Ascospores hyaline, cylindroid or allantoid, one celled, small.

Conidial state: *Endothiella* Sacc. (*Calopactis* Sydow & Sydow): Pycnidial cavities or locules formed in young perithecial stromata; conidiophores narrow, branched, conidiogenous cells phialidic; conidia hyaline, allantoid, one celled, small.

In branches of deciduous trees.

Type species: *Endothia gyrosa* (Schw. ex Fries) Fries.

Endothia in the present concept is limited to those species with brightly colored pseudoparenchymatous stromata, upright perithecia with central beaks, and allantoid, one-celled ascospores. The other species usually referred to *Endothia*, including *E. parasitica*, with prosenchymatous stroma, valsoid grouping of perithecia, and ellipsoid one-septate ascospores are segregated in the genus *Cryphonectria* Sacc. *Endothia tetraspora* Kobayashi (Trans. Mycol. Soc. Japan **6**: 99. 1965), with valsoid configuration of perithecia and large ellipsoidfusoid, one-celled ascospores belongs in *Cryptosporella*.

1. **Endothia gyrosa** (Schw. ex Fries) Fries, Summa Veget. Scand. p. 385. 1849. FIG. 64

Sphaeria gyrosa Schw. ex Fries, Syst. Mycol. **2**: 419. 1823.

Melogramma gyrosum (Schw. ex Fries) Tul. in Cooke, Ann. New York Acad. Sci. **1**: 185. 1878.

Stromata bright orange red to brownish externally, bright orange internally, 1-3 mm wide, 1-2 mm high; perithecia 150-300 μm diam, seated in base of stroma or in several layers, beaks elongate, erumpent at surface of stroma. Asci (16.5-)25-30 × (3.5-)6-7 μm. Ascospores (5.5-)7-11 × 1-3 μm.

Conidial state: *Endothiella gyrosa* Sacc.: formed in locules in young stroma; conidia 3-4 × 1-1.5 μm, hyaline, allantoid.

In branches of deciduous trees, especially species of *Quercus* but also *Acer*, *Fagus*, and *Liquidambar*, common in southeastern North America.

Material examined: *North America: Georgia:* GA 6572-6582 (GA).

Recently Roane et al. (**1974**) reported *E. gyrosa* as the causal organism of a blight of pin oak in Virginia.

In addition to the type species only a few taxa are recognized in the present narrow concept of *Endothia. Endothia singularis* Shear & Stevens (USDA Bull. **380**: 15. 1917), on species of *Quercus* in Colorado (Fungi Col. 4002, MASS), has large, brilliant red stromata and ascospores 5.5-11 × 1.5-3 μm. *Endothia viridistroma* Wehm. (Mycologia **28**: 38. 1936), on *Cercis canadensis* L. in Georgia (Athens, 30 Apr 1934, J.H. Miller, GA), has yellow green to dark or dull green stromata and ascospores 5-6 × 1-1.5 μm.

17. **Valseutypella** v. Höhnel, Ann. Mycol. **16**: 224. 1918.

Stromatic tissues pseudoparenchymatous, composed of sclerotial, thick-walled cells, stroma small, delimited by blackened exterior, interior yellowish; perithecia upright, few in stroma, beaks central to eccentric, erumpent separately. Asci ellipsoid, usually four spored. Ascospores hyaline, allantoid or ovoid or subgloboid, one celled.

Conidial state: *Cytospora*-like: Stromata similar to perithecial stromata, uniloculate; conidia hyaline, allantoid, one celled.

In branches of deciduous shrubs (*Rosa*), Europe, North America.

Type and single species *Valseutypella tristicha* (de Not.) v. Höhnel.

Hubbes (**1960b**) described and illustrated this fungus. The species produces both perithecial and conidial stromata in culture. Hubbes studied the physiology of *V. tristicha* and reported on variations among strains of the species.

1. **Valseutypella tristicha** (de Not.) v. Höhnel, Ann. Mycol. **16**: 224. 1918. FIG. 65

Diatrype tristicha de Not., Recl. Piren. Ital. p. 481. 1867.

Stromata ca. 1 mm diam, black externally, yellowish within; perithecia 300-350 μm diam, 250-300 μm high, beaks erumpent up to 100 μm beyond stroma surface, rounded at apex. Asci 55-70 × 6.5-10 μm, (one) four spored. Ascospores 11-24 × 3-4.5 μm, allantoid, at times ovoid or subgloboid and 9-16.5 × 4.5-7.5 μm, hyaline.

Conidia 3-5 × 1 μm.

In branches of *Rosa* spp., Europe, North America.

Material examined: *North America: Vermont:* Barr 4305b (MASS).

18. **Anisogramma** Theissen & Sydow, Ann. Mycol. **14**: 451. 1916.

Apioporthe v. Höhnel, Ber. Deutsch. Bot. Ges. **25**: 247. 1917; Sitzungsber. Kaiserl. Akad. Wiss. Math.-Naturwiss. Cl., Abt. 1, **126**: 382. 1917.

Stromatic tissues pseudoparenchymatous, stroma diatrypoid, seated on surface of wood and erumpent through periderm, delimited by darkened marginal zones, interior brown or gray; perithecia upright, in lower part of stroma, beaks central, not converging, only slightly protruding beyond surface, apices rounded or sulcate. Asci ellipsoid. Ascospores hyaline, ellipsoid, septum near lower end (apiosporous).

Conidial state: in *A. virgultorum* conidiophores arise from stroma surface, bear hyaline, ellipsoid, one-celled conidia (Müller and von Arx, **1962**).

In branches of deciduous trees.

Type species: *Anisogramma virgultorum* (Fries) Theissen & Sydow.

Both *Anisogramma* and *Apioporthe*, with type species *A. anomala*, were based upon the diatrypoid pseudoparenchymatous stroma and apiospores common to the two species. Petrak (**1934**) observed that *Anisogramma* was the earlier name for the genus. Wehmeyer (**1933**) did not consider *Anisogramma* in his treatment of genera related to *Diaporthe*, but did accept *Apioporthe*. He emended and expanded von Höhnel's concept of *Apioporthe* to include species which had a valsoid configuration of perithecia and stroma, and ascospores not truly apiosporous. Wehmeyer also included *Apioporthella* Petrak in synonymy; this apiosporous genus has a valsoid stroma. Müller in Müller and von Arx (**1962**) accepted *Anisogramma* in the original and in Petrak's sense, and reduced *Apioporthe* to synonymy.

Petrak (**1934**) reassigned the majority of species included by Wehmeyer (**1933**) in *Apioporthe* to positions in *Cryptodiaporthe* or in *Diaporthe*. Thus *Apioporthe corni*, *A. vepris*, *A. phomaspora* (and probably *A. apiospora*) belong in *Cryptodiaporthe*, while *A. lineariformis* and *A. utahensis* are in *Diaporthe*. *Apioporthe macrospora* Gilman, Tiffany & Lewis (Iowa State Coll. J. Sci. **330** 334. 1959) appears from the description and illustration to belong in *Cryptodiaporthe* or perhaps in *Allantoporthe*. The ascospores of this species are large, 30-36 × 9-10.5 μm.

1. **Anisogramma virgultorum** (Fries) Theissen & Sydow, Ann. Mycol. **14**: 451. 1916. FIG. 66

Sphaeria virgultorum Fries, Syst. Mycol. **2**: 351. 1823.

Hypoxylon virgultorum (Fries) Fries, Summa Veget. Scand. p. 384. 1849.

Dothidea virgultorum (Fries) Fckl., Symb. Mycol. 223. 1870.

Plowrightia virgultorum (Fries) Sacc., Syll. Fung. **2**: 636. 1883.

Apioporthe virgultorum (Fries) v. Höhnel, Sitzungsber. Kaiserl. Akad. Wiss. Math.-Naturwiss. Cl., Abt. 1, **126**: 382. 1917.

Apiospora rosenvingei Rostrup, Medd. om Grønl. **3**: 616. 1891.

Stromata 10-15 × 1-3 mm, up to 2 mm high; perithecia 350-500 μm diam, beaks to surface, apices rounded. Asci 45-52 × 10-13 μm. Ascospores 12-18 × 4.5-6.5 μm, lower cell 3-4.5 μm long, 3-3.5 μm wide.

In branches of *Betula* spp., Europe, North America.

Material examined: *North America: New Hampshire:* Chocorua, 12 Aug 1910, W.G. Farlow (NY).

2. **Anisogramma anomala** (Peck) Müller in Müller and von Arx, Beitr. Kryptogamenfl. Schweiz **11**(2): 769. 1962. FIG. 67

Diatrype anomala Peck, New York State Museum Rept. **28**: 72. 1876.

Cryptosporella anomala (Peck) Sacc., Syll. Fung. **1**: 470. 1882.

Cryptospora anomala (Peck) Ell. & Ev., North Amer. Pyreno. p. 531. 1892.

Apioporthe anomala (Peck) v. Höhnel, Sitzungsber. Kaiserl. Akad. Wiss., Math.-Naturwiss. Cl., Abt. 1, **126**: 381. 1917.

Stromata 2-5 × 2 mm, up to 2 mm high, surface grayish; perithecia 300-380 μm diam, compressed laterally and up to 1 mm high, beaks sulcate, black against stroma surface. Asci 35-40 × 9-11 μm. Ascospores 7.5-11 × 4-5 μm, lower cell very short, 1-2 μm long and wide.

In branches of *Corylus* spp., North America.

Material examined: *Maine:* Barr 5790a (MASS); *New York:* Albany, May, C.H. Peck (type, NYS); *Iowa:* Ellis N.A.F. 1185 (MASS).

19. **Diaporthella** Petrak, Ann. Mycol. **22**: 30. 1924.

Stromatic tissues pseudoparenchymatous, stroma diatrypoid, seated on wood surface and erumpent through periderm, blackened over surface and sides, brown within, some substrate cells incorporated at times; perithecia upright, in lower part of stroma or in several layers, beaks central, erumpent separately, not converging, opening at surface or extending far beyond surface. Asci ellipsoid. Ascospores fusoid ellipsoid, septum median, hyaline, appendages short and narrow when formed.

In branches of deciduous trees, Europe, North America.

Type species: *Diaporthella aristata* (Fries) Petrak.

Wehmeyer (**1933**) distinguished between *D. aristata* and *D. platasca*, saying that the latter species possessed larger ascospores and smaller stromata with a bluish cast on the surface. My examination of Peck's collection showed stromata variable in size and shape, with a blackened surface and no bluish cast, and ascospores 11-14 × 3-4 μm. The ascospores of *D. aristata* from European collections measure 12-13.5 × 2.5-3 μm according to Wehmeyer. These minor variations are not sufficient to maintain *D. platasca* as a separate species.

1. **Diaporthella aristata** (Fries) Petrak, Ann. Mycol. **22**: 30. 1924.
FIG. 68

Sphaeria aristata Fries, Syst. Mycol. **2**: 363. 1823.

Diatrype aristata (Fries) Fries, Summa Veget. Scand. 385. 1849.

Diaporthe aristata (Fries) Karst., Mycol. Fenn. **2**: 112. 1873.

Diatrype platasca Peck, New York State Museum Rep. **27**: 109. 1873.

Diaporthe platasca (Peck) Sacc., Syll. Fung. **1**: 613. 1882.

Diaporthella platasca (Peck) Wehm., The Genus *Diaporthe* p. 217. 1933.

Stromata 2-8 × 1-2 mm, rounded or elliptic; perithecia 385-640 μm diam, beaks elongate, projecting far beyond surface of stroma. Asci 35-52 × 9-12 μm. Ascospores (11-)12-17 × 2.5-4 μm, remnants of short appendages present.

In branches of *Betula* spp., Europe, North America.

Material examined: *North America: New York:* Nipple top Mt., Adirondacks, Aug 1873, C.H. Peck (NYS, type of *D. platasca*).

Diaporthella sphendamnina (Berk. & Curt.) Wehm. (The Genus *Diaporthe*, p. 216. 1933) on *Acer rubrum* L. has blackened marginal zones within the substrate in addition to the well-developed pseudoparenchymatous stroma, short beaks which are not erumpent beyond the stroma surface, and slightly narrower ascospores, 12-13.5 × 2.5-3 μm, in contrast to *D. aristata*.

20. **Sillia** Karsten, Mycol. Fenn. **2**: 20. 1873. (as "Hillia" by error.)

Stromatic tissues compact, prosenchymatous or pseudoparenchymatous, forming an erumpent pulvinate stroma with included periderm cells at times, surface blackened, interior bright ferruginous and pulverulent; perithecia upright, grouped or ±scattered in stroma, relatively few or numerous, beaks black and distinct in stromatic tissues, erumpent to or beyond stroma surface. Asci oblong. Ascospores hyaline, elongate fusoid, several septate.

In branches of deciduous trees.

Type species: *Sillia ferruginea* (Pers. ex Fries) Karsten.

1. **Sillia ferruginea** (Pers. ex Fries) Karsten, Mycol. Fenn **2**: 159. 1873. FIG. 69

Sphaeria ferruginea Pers. ex Fries, Mycol. **2**: 363. 1823.

Cryptospora ferruginea (Pers. ex Fries) Gilman, Tiffany & Lewis, Iowa State Coll. J. Sci. **33**: 328. 1959.

Stromata 1-4 mm diam or elongate and up to 4-6 × 2-3 mm, surface darkened or ferruginous at times, interior bright ferruginous; perithecia depressed, 330-440 μm diam, beaks broad, 165-190 μm, to surface or erumpent beyond stroma surface. Asci 60-80(-112) × 8-10(-16) μm. Ascospores (34-) 55-75 × 2.5-3 μm, hyaline, elongate fusoid, one, three, or five septate.

In branches of *Corylus* spp., Europe, North America. (Fries gave *Corylus, Populus, Sorbus, Betula*).

Material examined: *Europe:* GA 6310 (GA). *North America: Ontario:* GA 6309 (GA).

Only one North American specimen has been seen — the widely distributed collection made by R.F. Cain in 1931. This agrees with European material in all respects. Gilman, Tiffany and Lewis (**1959**) described two collections from Iowa.

II. Family **Valsaceae** Tulasne & Tulasne, Sel. Fung. Carp. **1**: 180. 1861, Barr emend.

Diaporthaceae v. Höhnel, Ber. Deutsch. Bot. Ges. **35**: 631. 1917 (sub Diaportheen).

Perithecia oblique or horizontal, beaks oblique or lateral, erumpent separately or converging through stromatic disc (2 subfamilies).

Key to Subfamilies and Tribes of Valsaceae

1. Perithecia single in substrate, stromatic tissues not formed .... Subfamily Plagiostomoideae, p. 100
One tribe, Plagiostomeae

1. Perithecia single or grouped in substrate, stromatic tissues prosenchymatous, forming a clypeus or blackened zone or stromatic disc at surface of substrate, at times delimited by blackened lateral and ventral marginal zones in substrate Subfamily Valsoideae . 2

2. Perithecia usually single in substrate, clypeus or blackened zone formed .... Tribe Linosporeae, p. 125

2. Perithecia usually grouped, covered by stromatic disc or blackened zone 3

3. Perithecia beneath stromatic disc, without blackened marginal zones .... Tribe Valseae, p. 131

3. Perithecia beneath blackened zones at surface of substrate, stromatic tissues often delimited by blackened lateral and ventral zones.......... Tribe Diaportheae, p. 146

II-1. Subfamily **Plagiostomoideae** Barr, subfam. nov.

Perithecia singulara, rostris erumpentis lateralis seorsim, stromata contexta nulla; in foliis et petiolis dicotyledonibus arboribus et frutibus vel herbarum caulis. Genus typicus: *Plagiostoma* Fckl. (one tribe: Plagiostomeae).

Key to Genera of Plagiostomoideae

1. Ascospores ellipsoid or fusoid or cylindroid, one septate........................ 2

1. Ascospores elongate cylindroid or acicular or fusoid, or filiform, one or usually several septate.. 3

2. Ascospores apiosporous .... 21. *Apioplagiostoma*

2. Ascospore septum medium or nearly so .... 22. *Plagiostoma*

3. Perithecial beaks narrow, short or elongate; in leaves of deciduous trees .... 23. *Pleuroceras*

3. Perithecial beaks relatively broad, short; in herbaceous stalks ..... 24. *Plagiosphaera*

The genus *Chadefaudiomyces* Kamat, Rao, Patil & Ullaso in Ullaso et al. (**1974**) described with horizontal perithecia, oblique beaks, and nonstromatic, evidently belongs in this subfamily. The sole species, *C. indicus* Kamat et al. develops in living leaves of *Celastrus paniculata* Willd. Ascospores are inversely apiosporous, i.e. the minute cell is cut off from the apical end, in contrast to the apiosporous condition in *Apioplagiostoma*.

II-1a. Tribe **Plagiostomeae**

21. **Apioplagiostoma** Barr, gen. nov.

Perithecia immersa, obliqua vel horizontalia, rostra excentrica vel lateralia. Asci ellipsoidei, unitunicati. Ascosporae hyalinae, ellipsoideae vel fusoideae, septatae prope basim. Species typica: *Plagiostoma populi* Cash & Waterman.

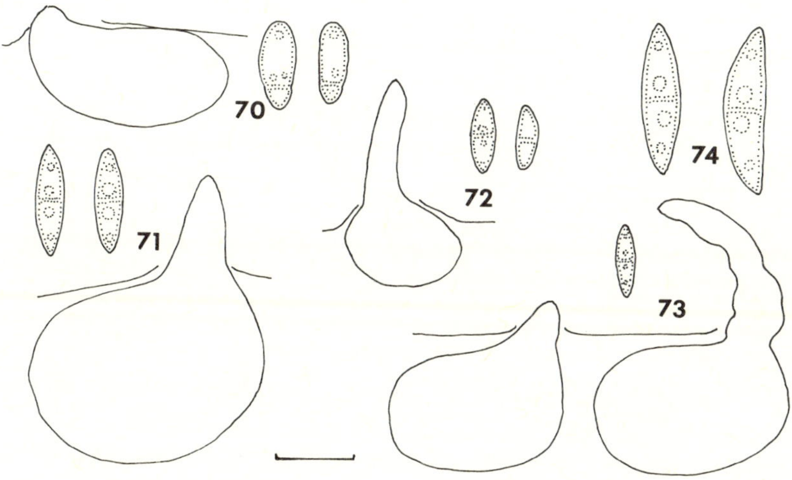

FIG. 70-74. FIG. 70. *Apioplagiostoma populi,* outline of perithecium, ascospores. FIGS. 71-74. Species of *Plagiostoma* section *Plagiostoma.* 71. *P. euphorbiae,* outline of perithecium, ascospores. 72. *P. conradii,* outline of perithecium, ascospores. 73. *P. devexa,* outline of perithecia, ascospore. 74. *P. lugubre,* ascospores. Scale =100 μm or 10 μm.

Stromatic tissues not formed; perithecia immersed, oblique to horizontal, beaks eccentric or lateral, relatively short and broad; peridium thin. Asci ellipsoid, bases loosening. Ascospores hyaline, ellipsoid or fusoid, straight or inequilateral, septum near base (apiosporous), lower cell one-third or less of total length, contents homogeneous or with several globules, wall smooth, overlapping uniseriate or biseriate in the ascus.

Conidial state not definitely known.

In leaves of deciduous trees and shrubs.

Type species: *Plagiostoma populi* Cash & Waterman.

This genus is erected to accomodate the apiosporous counterparts of *Plagiostoma*. It bears the same relationship to *Plagiostoma* as does *Apiognomonia* to *Gnomonia*. *Apioplagiostoma* is a replacement name for *Plagiostomella* v. Höhnel, Ann. Mycol. **17**: 131. 1919, non Ann. Mycol. **16**: 51. 1918, which is synonymous with *Plagiostoma* (see discussion under *Plagiostoma*).

1. **Apioplagiostoma populi** (Cash & Waterman) Barr, comb. nov.
FIG. 70

Plagiostoma populi Cash & Waterman, Mycologia **49**: 757. 1957.

Perithecia 200-320 μm diam, 100-150 μm high, depressed globose, horizontal, thickly scattered in leaf tissues, beak lateral, erumpent through either leaf surface, 40-70 μm high, (25-)50-80 μm wide; peridium 15-18 μm wide, light brown, beak wall blackened below pallid apex, some yellowish hyphae in leaf tissues. Asci 35-55 × 10-12 μm, (two) four (six or eight) spored. Ascospores 10-13 × 3.5-5(-6) μm, lower cell 2-3.5 μm long and wide.

In overwintered leaves of *Populus* spp., formed in brown spotted areas of leaves, North America.

Material examined: *Massachusetts:* Harvard Forest, Petersham, 11 May 1955, S.S. Pauley (type), 13 Sep 1954, 25 Oct 1954; 9 May 1956, A.M. Waterman, 18 Sep 1956, A.M. Waterman & K.F. Aldrich; Oct 1954, S.S. Pauley & A.M. Waterman; *New York:* Forest Home, Ithaca, 29 Mar 1934, G.E. Thompson (all BPI).

Cash and Waterman (**1957**) noted the presence of acervulus-like structure on the reddish brown leaf blades; conidia did not germinate, and their significance could not be determined. Ascospores did not germinate either on agar media or in distilled tap water.

Two European species are transferred to *Apioplagiostoma*.

2. **Apioplagiostoma carpinicola** (v. Höhnel) Barr, comb. nov.

Plagiostomella carpinicola v. Höhnel, Ann. Mycol. **16**: 52. 1918.

Gnomonia stahlii Klebahn, Haupt- und Nebenfruchtformen der Askomyzeten, p. 279. 1918.

Perithecia 110-200 μm diam, 80-110 μm high, depressed globose, horizontal in leaf tissues, beaks lateral, at times forming two which open to opposite leaf surfaces, or even a third at opposite end of perithecium, 40-100 μm long, up to 80 μm wide. Asci 42-56 × 8-13 μm. Ascospores (11-)14-21 × 3-4(-5) μm.

In overwintered leaves of *Carpinus betulus* L., Europe.

Material examined: *Europe:* Rehm, Ascomyceten 1425 (MICH, as *Laestadia carpinea*).

This species is the one von Höhnel proposed in **1919** to replace *P. petiolicola* (Fckl.) v. Höhnel as the type of *Plagiostomella* (**1918**). Von Höhnel cited Rehm's Ascomyceten 496 as well as 1425 for this species. However, Saccardo (**1882**) excluded n. 496 from *Laestadia carpinea* and in **1883** described the fungus represented by n. 496 as *Hypospila rehmii.* The ascospores are 16-20 × 3-3.5. μm, within the size range of *A. carpinicola*, but the septum is median, and von Arx's transfer of *H. rehmii* to *Plagiostoma* is maintained.

3. **Apioplagiostoma acerifera** (Cooke) Barr, comb. nov.

Sphaerella acerifera Cooke, Handbook, n. 2755. 1871.

Laestadia acerifera (Cooke) Sacc., Syll. Fung. **1**: 423. 1882.

Plagiostomella acerifera (Cooke) Petrak & Sydow, Ann. Mycol. **22**: 362. 1924.

Gnomonia acerina Starbäck, Bihang till Kongl. Svenska Vet.-Akad. Handl. 14, Afd. III, no. **5**: 17. 1889.

Pseudopeziza campestris Rehm, Ber. Bayr. Bot. Ges. München **13**: 167. 1912.

Drepanopeziza campestris (Rehm) Jaap, Fung. sel. exs. n. 556.

Plagiostomella campestris (Rehm) v. Höhnel, Sitzungsber. Kaiserl. Akad. Wiss., Math.-Naturwiss. Cl., Abt. 1, **127**: 627. 1918.

Perithecia 220-250 μm diam, beaks short, lateral, 100 μm long, 70-80 μm wide. Asci 65-75 × 12-18 μm. Ascospores 12-13.5(-18) × 4.5-5 (-6) μm, lower cell 3-4 μm long.

In overwintered leaves of *Acer* spp., Europe.

Material examined: *Europe:* Sydow, Mycol. germ. 587 (TRTC, as *Gnomonia acerina*).

22. **Plagiostoma** Fuckel, Symb. Mycol. p. 118. 1870, emend von Arx (**1951**)

Laestadia Auersw., Hedwigia **8**: 177. 1869, non Kunth 1832.

Guignardia Viala & Ravaz, Bull. Soc. Mycol. France **9**: 63. 1892.

Gnomonina v. Höhnel, Ann. Mycol. **16**: 46. 1918.

Plagiostomella v. Höhnel, Ann. Mycol. **16**: 51. 1918.

Chalcosphaeria v. Höhnel, Ann. Mycol. **16**: 97. 1918.

Stromatic tissues not formed; perithecia immersed in host tissues, oblique or horizontal, depressed globose or elliptic, beaks usually erumpent hypophyllously, epiphyllously at times, eccentric or lateral, occasionally lacking and periphysate pore formed, short or elongate, usually relatively broad; peridium narrow, of compressed layers of cells, with sparse hyphae into substrate, at times surrounded by layer of crystalline material. Asci oblong ellipsoid. Ascospores hyaline, rarely brownish in age, ellipsoid, fusoid, occasionally nearly cylindroid, usually straight to inequilateral, ends rounded or pointed, one septate, septum median or supramedian or submedian, usually with two globules per cell, wall smooth, occasionally becoming roughened in age, at times with terminal appendages which are short and pulvinate or elongate and strap-like or narrow and setose, overlapping biseriate to triseriate in the ascus, rarely in a fascicle.

Conidial state not definitely known.

Usually saprobic and maturing in overwintered leaves and petioles of deciduous trees or herbaceous plant stalks.

Type species: *Plagiostoma euphorbiae* (Fck.) Fckl.

Plagiostoma was erected for species in which the perithecia were horizontal in respect to the substrate surface and opened by a lateral beak. Most authors have not differentiated a genus based upon this character, and have included the species of *Plagiostoma* under *Gnomonia*, although v. Höhnel (**1917**) did separate the two. Von Arx (**1951**) investigated several species and redefined *Plagiostoma* as closely related to *Gnomonia*, but differing in the horizontal position of perithecia and lateral beaks versus upright perithecia and central beaks. Von Arx included the type species of *Laestadia* and *Gnomonina* (*Sphaeria alnea* Fries) and of *Chalcosphaeria* (*Sphaeria pustula* Pers. ex Fries) in *Plagiostoma. Guignardia* had been intended as a replacement for *Laestadia* and is a nomenclatural synonym. *Plagiostomella* was based upon *P. petiolicola* Fckl. (v. Höhnel, **1918**), as the apiosporous counterpart of *Plagiostoma.* However, Klebahn (**1918**) pointed out the true nature of the ascospores, which had been reported erroneously as apiosporous. Von Höhnel (**1919**) recognized then that *P. petiolicola* was foreign to his concept of *Plagiostomella* and proposed that the second species, *P. carpinicola* v. Höhnel be taken as type of the genus. Although the change in typification has been accepted by some authors (von Arx, **1951**, Müller and von Arx **1962**, **1973**), such change is unjustifiable under the International Code of Botanical Nomenclature (Cash and Waterman, **1957**). Thus, *Plagiostomella* must be considered typified by the original intent, i.e., *P. petiolicola*, and is a synonym of *Plagiostoma. Apioplagiostoma* is proposed in this study to accommodate apiosporous species otherwise closely related to *Plagiostoma.*

Key to Sections and Species of *Plagiostoma*

1. Ascospores narrow (length:width ±5-6:1), septum median Section 5. ***Angustispora*** . 2
1. Ascospores relatively broad (length:width ±3-4:1). 4
   2. Peridium reddish brown, beak yellow to reddish brown; ascospores (15.5-)18-26 (-32) × (1-)2-2.5(-3) μm, appendages setose. . . . . 13. ***P. magnoliae***
   2. Peridium and beak dark brown; ascospores wider, 2.5-4 μm, appendages straplike or blunt 3
3. Ascospores 18-27(-30) × 2.5-4 μm . . . . 11. ***P. campylostyla*** var. ***campylostyla***
3. Ascospores 27.5-37.5 × 3-3.5 μm . . . . 12. ***P. campylostyla*** var. ***mirabilis***
   4. Ascospores fusoid, tapering to ±pointed ends, appendages not usually formed, setose if formed 5
   4. Ascospores ellipsoid, tapering to rounded ends, appendages blunt or pulvinate. 9
5. Ascospore septum median. . . . . Section 1. ***Plagiostoma*** . 6
5. Ascospore septum supramedian. . . . . Section 2. ***Plagiostomella***
   A single species, in Europe . . . . ***P. petiolicola***
   6. Ascospores 8-11 × 2-4 μm 7
   6. Ascospores larger 8
7. Perithecia small, 100-160 μm diam; ascospores 3-4 μm wide. . . . . 2. ***P. conradii***
7. Perithecia larger, 165-300 μm diam; ascospores 2-3 μm wide . . . . 3. ***P. devexa***
   8. Ascospores (9-)12-16(-18) × 3-4.5 μm. . . . . 1. ***P. euphorbiae***
   8. Ascospores 17-24 × 5.5-7 μm . . . . 4. ***P. lugubre***

9. Ascospore septum median .... Section 3. *Guignardia* . 10

9. Ascospore septum submedian .... Section 4. *Disparicellula* . 13

10. Ascospores 8-12(-15) × 2-3 μm, beaks short and broad .... 8. *P. bavarica*

10. Ascospores broader, 3.5-6.5(-10) μm, beaks usually elongate................ 11

11. Ascospores 26-36 × 5.5-6.5(-10) μm; beaks broad .... 7. *P. micromegala*

11. Ascospores (10-)15-20(-25) × 3.5-6 μm; beaks narrow......................... 12

12. Perithecia small, 200-300 μm diam...................... 5. *P. alnea* var. *alnea*

12. Perithecia larger, 385-492 μm diam.................. 6. *P. alnea* var. *betulina*

13. Ascospores 6-10 × 1.5-3 μm, beaks narrow .... 10. *P. tormentillae*

13. Ascospores 15.5-26(-33) × 4-5.5(-6.5) μm, beaks broad .... 9. *P. acerophila*

Plagiostoma section 1. **Plagiostoma**

Ascospores fusoid, tapering to ±pointed ends, septum median, appendages usually not formed, biseriate in the ascus; beaks narrow, usually short. Type species: *Plagiostoma euphorbiae* (Fckl.) Fckl.

1. **Plagiostoma euphorbiae** (Fckl.) Fckl. Symb. Mycol. p. 118. 1870.
FIG. 71

Sphaeria euphorbiae Fckl., Enum. Fung. Nassov. p. 69. 1868.

Gnomonia euphorbiae (Fckl.) Sacc., Michelia **2**: 312.

Gnomoniella euphorbiae (Fckl.) Sacc., Syll. Fung. **1**: 418. 1882.

Perithecia scattered to gregarious, immersed, depressed globose, 275-400 μm diam, beaks lateral or eccentric, broad, short, and ±conic or elongate, 65-400 μm or longer, 65 μm or more wide near base, tapering slightly to apex. Asci 40-64 × 7.5-10 μm. Ascospores (9-)12-16(-18) × 3-4.5 μm, septum median, ellipsoid fusoid, tapering to pointed ends, overlapping biseriate in the ascus.

Conidial state (according to Fuckel): *Discosia euphorbiae* Fckl.

In dead stalks of *Euphorbia* spp., Europe.

Material examined: *Europe:* Smarods, Fungi latvici exs. 720 (TRTC).

Probable additional synonyms are:

Gnomonia amygdalinae Fckl., Symb. Mycol. p. 121. 1870.

Gnomoniella amygdalinae (Fckl.) Sacc., Syll. Fung. **1**: 418. 1882.

Gnomonia devexa Auersw., Mycol. Europ. **5/6**: 23. 1869, pro parte sec Winter (**1887**).

Gnomoniella devexa var. *media* Sacc., Syll. Fung. **1**: 417. 1882.

Gnomonia euphorbiacea Sacc. & Briard, Rev. Mycol. 7(28): 208. 1885.

Gnomonia euphorbiacea f. *wulfenii* Rehm in Jaap, Ann. Mycol. **14**: 17. 1916.

Gnomonia tithymalina Briard & Sacc., Rev. Mycol. 7(28): 209. 1885.

This species has not yet been found in North America. It is reported here to illustrate the typical section, which does have other North American taxa.

2. **Plagiostoma conradii** (Ellis) Barr, comb. nov. FIG. 72

Diaporthe conradii Ellis, Amer. Nat. **17**: 816. 1883.

Gnomonia conradii (Ellis) Wehm. sec. Dearness & House, New York State Museum Circ. **24**: 38. 1940.

Perithecia 100-160 μm diam, up to 117 μm high, depressed globose, scattered, beak lateral, 100-200 μm long, 45-60 μm near base, tapering to 30-40 μm near apex. Asci 30-48 × 6-7 μm. Ascospores 9-11 × 3-4 μm, fusoid, overlapping biseriate in the ascus.

In dead branches of *Hudsonia ericoides* L., *H. tomentosa* Nutt., eastern coastal North America.

Material examined: *Massachusetts:* Wellfleet, 16 Aug 1932, A.S. Goodale & W.C. Markert (MASS); Falmouth, 23 Sep 1932, D. Potter (MASS); *New Jersey:* Newfield, Nov 1882, J.B. Ellis (type, and Ellis N.A.F. 1193, MASS).

Ellis and Everhart (**1892**) noted that the specific epithet was the result of mistaken identification of the host plant as *Corema conradii* Torr. in the Empetraceae, rather than *Hudsonia tomentosa* in the Cistaceae. Wehmeyer (**1933**) pointed out that the fungus belonged in *Gnomonia* rather than in *Diaporthe* but he did not transfer the species.

3. **Plagiostoma devexa** (Desm.) Fckl., Symb. Mycol. p. 119. 1870. FIG. 73

Sphaeria devexa Desm., Crypt. France Ed. II, no. 367.

Gnomoniella devexa (Desm.) Sacc., Syll. Fung. **1**: 417. 1882.

Gnomoniopsis devexa (Desm.) Moesz & Smarods, Bot. Közl. **38**: 68. 1941.

Sphaeria euphorbiae f. *polygoni* Fckl., Fungi rhen. 864.

Gnomonia obliqua Auersw., Leipz. Tausch-Verein, p. 4. 1870.

Sphaeria eccentrica Cooke & Peck, New York State Museum Rep. **25**: 105. 1873.

Gnomoniella eccentrica (Cooke & Peck) Sacc., Syll. Fung. **1**: 418. 1882.

Diaporthe sechalinensis Sacc., Atti Congr. Bot. Palermo **1902**: 52. 1902.

Ceriosporella polygoni Smith & Ramsb., Trans. Brit. Mycol. Soc. **4**: 325. 1914.

Perithecia immersed, depressed globose, 165-300 μm diam, 105-150 (-200) μm high, beaks eccentric to lateral, 90-210(-315) μm long, 45-67 μm wide near base, tapering to 30 μm near apex. Asci 29-38 × 5.5-7 μm, apical annulus minute. Ascospores 8-10 × 2-3 μm, ellipsoid fusoid, septum median, no appendages seen, (but setose ones have been described), overlapping biseriate in the ascus.

In overwintered stalks of *Polygonum* spp., Europe, North America; also reported on *Rumex* spp.

Material examined: *Europe:* Herb. Barbey-Boissier 82 (TRTC); Rylland, Wiborg, Denmark, 28 Jun 1904, J. Lind (DAOM 115561). *North America: New York:* Albany, Jun 1872, C.H. Peck (type of *Sphaeria eccentrica*, NYS, NY); Portville, Sep, C.H. Peck (NYS); Buffalo, G.W. Clinton (BPI, NYS).

Pirozynski and Morgan-Jones (**1968**) observed that *Ceriosporella polygoni* and *Diaporthe sechalinensis* were identical with *P. devexa*. The North American collections of *Sphaeria eccentrica* are in accord with European ones of *P. devexa*.

4. **Plagiostoma lugubre** (Karst.) Bolay, Ber. Schweiz. Bot. Ges. **81**: 436. 1972. FIG. 74

Gnomonia lugubris Karst., Mycol. Fenn. **2**: 121. 1873.

Gnomoniella lugubris (Karst.) Sacc., Syll. Fung. **1**: 415. 1882.

Perithecia immersed, depressed globose, 240-270 μm diam, 180-200 μm high; beak lacking at times, usually elongate, up to 270 μm long, ca. 60 μm wide near base. Asci 47.5-70 × 9.5-15 μm. Ascospores 17-24 × 5.5-7 μm, ellipsoid fusoid, often inequilateral, overlapping biseriate in the ascus.

In overwintered leaves and petioles of *Potentilla palustris* (L.) Scop., Europe, western North America.

Material examined: *North America: British Columbia:* Beaver Lake, Stanley Park, Vancouver, 29 Mar 1952, M.E. Barr (UBC, DAOM).

Plagiostoma section 2. **Plagiostomella** (v. Höhnel) Barr, stat. nov.

Ascospores fusoid, tapering to ±pointed ends, septum supramedian, appendages not formed, biseriate in the ascus; beaks narrow. Type species: *Plagiostoma petiolicola* Fckl. (Symb. Mycol. p. 119. 1870).

This species is the only one presently known in section *Plagiostomella.* It occurs on *Tilia* leaves in Europe, has ascospores 12-15 × 3-3.5 μm, with the upper cell about one-third the length and usually narrower than the lower cell. *Plagiostoma petiolicola* has not yet been collected in North America.

Plagiostoma section 3. **Guignardia** (Viala & Ravaz) Barr, stat. nov.

Ascospores ellipsoid, tapering to rounded ends, septum median, appendages broad and short when formed, biseriate to obliquely fasciculate in the ascus; beaks narrow, short to elongate, not formed at times. Type species: *Plagiostoma alnea* (Fries) von Arx.

In this section, two European species not yet discovered in North America should be mentioned. *Plagiostoma rehmii* (Sacc.) von Arx (Antonie v. Leeuwenh. **17**: 264. 1951), on leaves of *Carpinus*, has small perithecia and ascospores 16-20 × 3-3.5(-4.5) μm, narrower than those of *P. alnea. Plagiostoma pustula* (Pers. ex Fries) von Arx (Antonie v. Leeuwenh. **17**: 264. 1951) develops in leaves of *Quercus* and is distinctive by the crustose layer of granular material, shining light brown externally and whitish within, which surrounds each perithecium. On the basis of this character, *P. pustula* was segregated into the genus *Chalcosphaeria* by v. Höhnel (**1918**). As von Arx (**1951**) observed, the character is not sufficient for generic separation. The granular layer which forms a collar around the base of beaks in several species of *Gnomonia* would appear to be homologous. The ascospores are 17-25 × 4-6 μm and bear a short broad appendage at each end. Morgan-Jones (**1959**) observed that development of the perithecia was similar to that of the species of *Gnomonia* (*setacea*, *gnomon*, *intermedia*) that he studied.

5. **Plagiostoma alnea** (Fries) von Arx, Antonie van Leeuwenhoek **17**: 264 1951. var. **alnea**. FIG. 75

Sphaeria alnea Fries, Syst. Mycol. **2**: 520. 1823.

Sphaerella alnea (Fries) Auersw., Mycol. Europ. **5/6**: pl. 2, fig. 15. 1869.

Laestadia alnea (Fries) Auersw., Hedwigia **8**: 177. 1869.

Guignardia alnea (Fries) Schroet., Pilze Schlesiens **2**: 330. 1894.

Gnomonina alnea (Fries) v. Höhnel, Ber. Deutsch. Bot. Ges. **35**: 628. 1917; Ann. Mycol. **16**: 47. 1918.

Gnomonia perversa Rehm, Hedwigia **24**: 70. 1885.

Gnomonia vleugelii Kleb., Haupt- und Nebenfruchtformen der Askomyzeten, p. 251. 1918.

Plagiostoma vleugelii (Kleb.) v. Höhnel, Hedwigia **62**: 49. 1921.

Perithecia immersed in leaf tissues or petioles, rounded or elliptic, depressed, (110-)200-300 μm diam, 135-245 μm high, oblique or horizontal, beaks lacking, short, or elongate, up to 500 μm long, 30-80 μm wide near base, eccentric or lateral, usually curved, erumpent through either leaf surface but mostly hypophyllous; peridium 10-15 μm wide. Asci (46-)60-90 × 9-14 μm, eight spored. Ascospores (10-)15-20(-25) × 3.5-6 μm, ellipsoid, ends rounded, straight to inequilateral, septum median, appendages short pulvinate, 2-4 μm wide and long, overlapping biseriate in the ascus.

In overwintered leaves and petioles of *Alnus* spp., Europe, North America.

Material examined: *Europe:* Krieger, Fungi sax. 77 (as *G. perversa*, MICH, NY); Rabenhorst-Winter F. eur. n. 3455 (as *G. perversa*, NY); Rehm Ascom. n. 795 (type of *Gnomonia perversa*, NY); Sydow, Mycoth. March 2953 (as *G. "subversa"*, NY); Vestergren Micromyc. rar. sel. 1144 (on type specimen of *Gnomoniella alnobetulae*). *North America: British Columbia:* Barr 5582, 5775 (MASS); *Vermont:* Barr 2986 (MASS); *Colorado:* Grand Mesa, Summer 1930, R.W. Davidson (BPI).

Klebahn (**1918**) germinated ascospores, the hyphae from which after several months produced mature perithecia. No conidial state was formed in culture nor on leaves inoculated with ascospores.

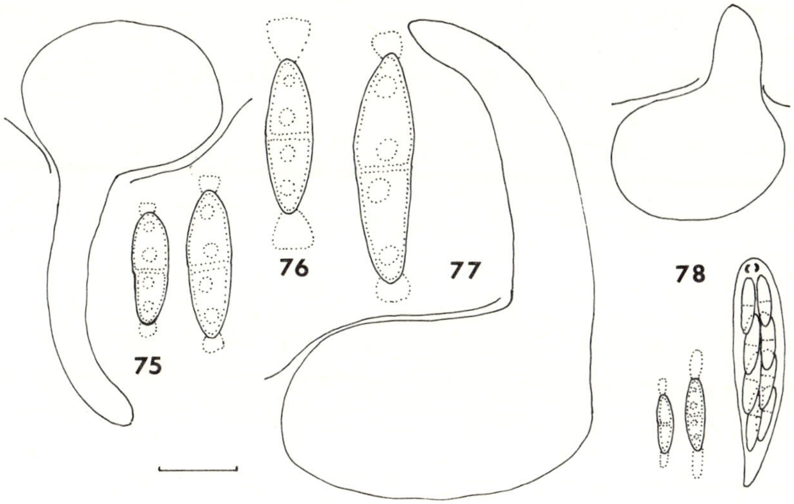

FIGS. 75-78. Species of *Plagiostoma* section *Guignardia*. 75. ***P. alnea*** var. ***alnea,*** outline of perithecium, ascospores. 76. ***P. alnea*** var. ***betulina,*** ascospore. 77. ***P. micromegala,*** outline of perithecium, ascospore. 78. ***P. bavarica,*** outline of perithecium, ascus, ascospores. Scale =100 μm or 10 μm.

6. **Plagiostoma alnea** var. **betulina** Barr, var. nov. FIG. 76

Similis vel var. *alneae* sed perithecia ampliora et appendiculae magniores. Holotypus in foliis *Betulae glandulosae* Michx., "Haines Highway, British Columbia, 17 Jul 1959," a W.G. Ziller lecto, (DAOM 71511).

Perithecia 385-495 μm diam, 220-275 μm high, depressed globose, beaks erumpent from either leaf surface, 495-550 μm long, 52-65 μm wide near base, enlarged below apex to 70-78 μm wide. Asci 50-70 × 14-20 μm, apical annulus 3 μm high. Ascospores (18-)20-23 × 4.5-5.5 (-6) μm, ellipsoid, appendages 5 μm high, 5-7 μm wide, in oblique fascicle in the ascus.

In leaves of *Betula glandulosa* Michx.

Known only from the type collection.

D. B. O. Savile had determined this collection as *Gnomonia campylostyla,* but that species has narrower, more cylindroid ascospores with strap-like appendages.

7. **Plagiostoma micromegala** (Ell. & Ev.) Barr, comb. nov. FIG. 77

Diaporthe micromegala Ell. & Ev., Proc. Acad. Nat. Sci. Phila. for **1893**: 449. 1894.

Perithecia 320-480 μm diam, 200-330 μm high, depressed globose, beaks 495-600 μm long, 75-130 μm wide near base, tapering to 60-78 μm near apex. Asci 67-82 × 16.5-24 μm, apical annulus 5-6 μm high. Ascospores 26-36 × 5.5-10 μm, ellipsoid cylindroid, ends rounded, septum median, appendages broad, 2-6.5 μm high, 3-4 μm wide.

In overwintered leaflets and rachises of *Carya* spp., North America.

Material examined: *Delaware:* Wilmington, 24 Jul 1893, A. Commons (NY, type); *Georgia:* GA 6616, 6617, 6666, 6667 (GA).

Wehmeyer (**1933**) excluded this species from *Diaporthe* because of scattered perithecia and no blackened zones in the substrate. He noted that it was a species of *Gnomonia* but made no combination. The ascospore measurements which he gave (38-40 × 6-7 μm) are longer than those I obtained, and than those given in Ellis's notes with the type specimen. Perhaps he included the length of appendages in the measurements, for he did not mention appendages.

The type specimen was described fron unknown dead stems; on the packet of type material is marked "*Desmodium?*." This material agrees so well with the ample specimens on *Carya* rachises from Georgia that they must be considered as one species. The sole difference noted was in arrangement of ascospores in the ascus — oblique in Delaware material, in a fascicle in the Georgia material.

8. **Plagiostoma bavarica** (Rehm) Barr, comb. nov. FIG. 78

Hypospila bavarica Rehm, Ann. Mycol. **6**: 322. 1908.

Perithecia 220-275 μm diam, depressed globose, epiphyllous in leaf blade and in petioles, separate or two together, beaks lateral, short, 104-165 μm long, 65-80 μm wide, tapering to apex. Asci 25-35 × 5-7 μm. Ascospores 8-12 × 2-2.5 μm, ellipsoid, septum ±median, appendages blunt, 2-5 μm long, biseriate in the ascus.

In overwintered leaves of *Acer pseudoplatanus* L., Europe, *A. saccharum* Marsh., North America.

Material examined: *North America: Georgia:* GA 6701 (GA).

A single leaf comprises the above collection. The fungus is well developed and its characters agree with the description of *Hypospila bavarica.* Rehm's measurements of ascospores, 12-15 × 3-3.5 μm, are slightly larger, but the range of sizes could well encompass the smaller ones of the Georgia collection. Rehm's species is quite different from *Gnomoniella bavarica* Kirschst. (Ann. Mycol. **33**: 214. 1935), which is a synonym of *Physalospora vitis-idaeae* Rehm in Krieger (Barr, **1970**).

Plagiostoma section 4. **Disparicellula** Barr, sectio nov.

Ascosporae ellipsoideae, cellulis disparibus, inferiore brevioribus. Species typica: *Plagiostoma acerophila* (Dearn. & House) Barr.

Ascospores ellipsoid, tapering to rounded ends, septum submedian, appendages short and blunt or somewhat elongate, biseriate or in overlapping fascicle in the ascus; beaks narrow or broad. Type species: *Plagiostoma acerophila* (Dearn. & House) Barr.

In Europe, **Plagiostoma robertiani** (Petrak) Barr, comb. nov. (Basionym: *Gnomonia robertiani* Petrak, Ann. Mycol. **23**: 122. 1925) is another species belonging in this section. Data on the species, from study of Petrak's Mycotheca generalis 80 (NY), on *Geranium robertianum* L., M. Weisskirchen, Moravia, May 1924: Perithecia 380-450 μm diam, 200-250 μm high, beak lateral, up to 600 μm long, 60-80 μm wide; asci 55-60 × 10-15.5 μm, apical annulus 3.5 μm high; ascospores 14.5-19.5 × 4.5-6 μm, ellipsoid obovoid, ends rounded, septum submedian, appendages 2-3.5 μm wide and high. *Gnomonia borealis* Schroet. (Jahres-Ber. Schles. Ges. f. vaterl. Cultur **65**: 275. 1888; *Gnomoniella borealis* (Schroet.) Kirschst. Ann. Mycol. **33**: 227. 1935) according to the description may be identical with *P. robertiani.* So also may be *Gnomoniella kriegeriana* Rehm (Ann. Mycol. **13**: 2. 1915), although this fungus was described with smaller perithecia and one-celled ascospores.

9. **Plagiostoma acerophila** (Dearn. & House) Barr, comb. nov. FIG. 79

Gnomoniopsis acerophila Dearness & House, Bull. New York State Museum **233-234**: 36. 1921.

Gnomonia acerophila (Dearn. & House) Barr in Bigelow & Barr, Rhodora **62**: 196. 1960.

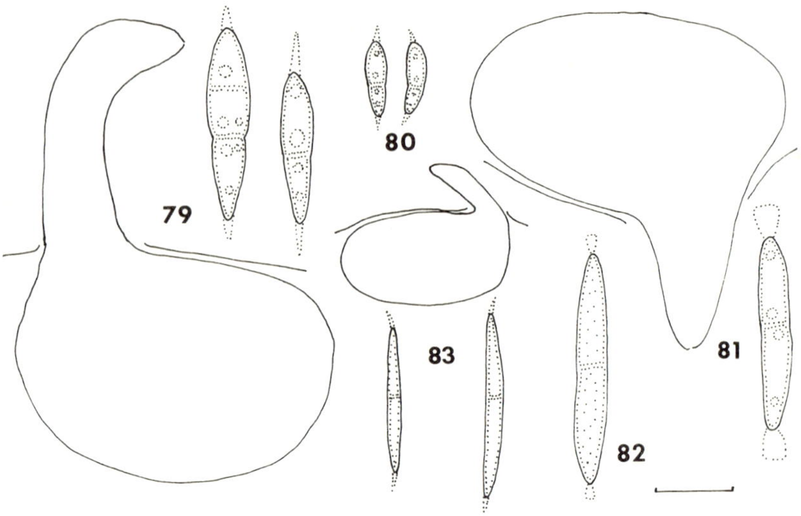

FIGS. 79-83. Species of ***Plagiostoma*** section ***Disparicellula*** (79, 80), section ***Angustisporae*** (81-83). 79. ***P. acerophila,*** outline of perithecium, ascospores. 80. ***P. tormentillae,*** ascospores. 81. ***P. campylostyla*** var. ***campylostyla,*** outline of perithcium, ascospore. 82. ***P. campylostyla*** var. ***mirabilis,*** ascospore. 83. ***P. magnoliae,*** outline of perithecium, ascospores. Scale =100 μm or 10 μm.

Perithecia 420-600 μm diam, 250-400 μm high, beak lateral, 300-1000 μm long or up to 2 mm long, 100-130 μm wide near base, tapering to 60-70 μm wide near apex. Asci 50-89 × 12-17.5 μm, apical annulus 3 μm high. Ascospores 15.5-26(-33) × 4-5.5(-6.5) μm, ellipsoid fusoid, septum submedian, occasionally a second septum formed in upper cell in age, becoming brownish in age, appendages (2-)6-9(-18) μm long, 1-3 μm wide.

In petioles, main veins of overwintered leaves of *Acer rubrum* L., less frequently *A. pensylvanicum* L. and *A. spicatum* Lam., North America.

Material examined: *Quebec:* Barr 1864 (MASS); *Ontario:* D. Malloch 648 (MASS; TRTC 50416); *Maine:* Barr 5886a, 3420b (MASS); *New Hampshire:* Barr 3816a (MASS); *Vermont:* Barr 4222 (MASS); *Massachusetts:* Barr 5596 (MASS); *New York:* Barr 1888a (MASS); Cain 648 (TRTC); Oneida, 15 May 1918, H.D. House (type, NYS, NY); *New Jersey:* Ell. & Ev. N.A.F. 2139 (as *Gnomonia emarginata* Fckl. MASS, MICH); Newfield, Jun 1888, J.B. Ellis (NY, as *Gnomonia emarginata* Fckl); *Michigan:* Barr 886, 5482a (MASS); *Louisiana:* Barr 6339 (MASS).

Plagiostoma acerophila was described as a species of *Gnomoniopsis* because of the occasional occurrence of secondary septa in age. When Ellis and Everhart (1888) described their collection (NAF 2139), they referred it to *Gnomonia emarginata* "for the present," although they noted discrepancy in ascospore size and shape. This species usually occurs on the bases of petioles while *Pleuroceras tenella* occupies the upper petiole.

10. **Plagiostoma tormentillae** (Lind) Bolay, Ber. Schweiz. Bot. Ges. **81**: 436. 1872. FIG. 80

Gnomoniella tormentillae Lind, Bot. Tidsskrift **41**: 217. 1928.

Perithecia 200-300 μm diam, 180-250 μm high, depressed, beaks 150-180 μm long, 40-45 μm wide near base. Asci 35-45 × 6-6.5 μm. Ascospores 6-10 × 1.5-3 μm, ellipsoid fusoid, septum submedian, appendages ca. 2 μm long, 1 μm wide, blunt.

In overwintered petioles, veins and stalks of *Potentilla* spp., Europe, North America.

Material examined: *Massachusetts:* Barr 4818 (MASS).

This collection agrees with Bolay's description of the species from European collections.

Plagiostoma section 5. **Angustisporae** Barr, sectio nov.

Ascosporae angustae, cylindroidae fusoideae, septum medium. Species typica: ***Plagiostoma campylostyla*** (Auersw.) Barr.

Ascospores narrow, cylindroid fusoid, septum median, appendages broad and short or elongate, or narrow and setose, in a fascicle or overlapping in the ascus; beaks narrow.

The European **Plagiostoma inclinata** (Desm.) Barr, comb. nov. (Basionym: *Sphaeria inclinata* Desm., Ann. Sci. Nat. Bot. III, **16**: 315. 1851; *Gnomonia inclinata* (Desm.) Auersw. in Rabenh., Mycol. Europ. **5/6**: 27. 1869) on *Acer campestris* L., belongs in this section. From an authentic specimen, Desmazieres Plantes Crypt. Fr. 2081, NY, the following data were obtained: Perithecia 230-300 μm diam, 100-130 μm high, hypophyllous, beaks lateral, 100-180 μm long, 30-40 μm wide near base; asci 48-55 × 6-10 μm; ascospores 14.5-18.5 × 2-3(-3.5) μm, appendages elongate, up to 22 μm long, 1-1.5(-2) μm wide, overlapping biseriate in the ascus.

11. **Plagiostoma campylostyla** (Auersw.) Barr, comb. nov. var. **campylostyla** FIG. 81

Gnomonia campylostyla Auersw. in Rabenh., Mycol. Europ. **5/6**: 25. 1869.

Perithecia 385-550 μm diam, 280-350 μm high, depressed globose, beaks hypophyllous, eccentric to lateral, rarely lacking, short, 130-165 μm long or elongate, 460-616(-1300) μm long, 52-72 μm wide near base. Asci 48-78 × 9-13.5 μm. Ascospores 18-27(-30) × 2.5-4 μ, occasionally with one or two additional septa, appendages 2.5-4.5 μm long, 2-3 μm wide, in two fascicles or obliquely fasciculate in the ascus.

In overwintered leaves of *Betula* spp., Europe, North America.

Material examined: *North America: Quebec:* Barr 1900, 1901, 2039 (DAOM, MASS); *Ontario:* TRTC 12263, 35824 (TRTC); *Maine:* Barr 1660b, (MASS); *New Hampshire:* Barr 3817 (MASS); *Vermont:* Barr 2987, 4202 (MASS); *Massachusetts:* Barr 2315, 6058, 6063 (MASS); *Michigan:* Barr 952 (MASS); *Tennessee*: Greenbrier Cove, Great Smoky Mts. Nat. Park, 13 Apr 1946, L.R. Hesler (TENN 17421).

Gnomonia emarginata Fuckel (Symb. Mycol. p. 102. 1870) (*Gnomoniella emarginata* (Fckl.) Sacc., Syll. Fung. **1**: 414. 1882) may be identical with Auerswald's species. The narrow lateral beaks suggest this; Fuckel gave no sizes for asci and ascospores. In their discussion of *G. emarginata*, Ellis and Everhart (**1888**) noted that most exsiccati specimens were sterile, but that Rabenhorst-Winter n. 2756 had lateral beaks, asci 35-40 × 7 μm, and ascospores 20-22 × 2-2.5 μm. Except for the narrower ascospores, such a description would fit within the range of *P. campylostyla.*

Gnomonia betulina Vleugel (Svensk Bot. Tids. **11**: 305. 1917; *G. campylostyla* Auersw. f. *odoratae* Vleugel, Svensk Bot. Tids. **5**: 344. 1911) is also probably identical with *P. campylostyla.* Vleugel described it on leaves of *Betula odorata*, while *P. campylostyla* occurs on *Betula verrucosa, B. alba, B. nana.*

Klebahn (**1918**) obtained ascospore germination and formation of a mycelial colony. Neither perithecia nor conidia were produced in culture. Attempts to infect living leaves were negative.

12. **Plagiostoma campylostyla** var. **mirabilis** (Peck) Barr, comb. nov. FIG. 82

Sphaeria mirabilis Peck, New York State Museum Rep. for 1874, **28**: 80. "1875" 1876.

Gnomoniella mirabilis (Peck) Sacc., Syll. Fung. **1**: 414. 1882.

Perithecia 450-615 μm diam, 230-245 μm high, depressed, beaks up to 700-920 μm long, 50-75 μm wide and tapering to apex. Asci 52-67 × 10.5 μm. Ascospores 27.5-37.5 × 3-3.5 μm, appendages 2.5-6 μm long, 1-2 μm wide, in oblique fascicle in the ascus.

In overwintered leaves of *Betula lutea* Michx. f., North America.

Material examined: *New York:* Bethlehem, Jun, C.H. Peck (type, NYS, NY).

The host, habit, and structure of this variety are all similar to var. *campylostyla*. Only greater sizes distinguish Peck's collection, which is regarded as a large variety of *P. campylostyla*.

13. **Plagiostoma magnoliae** (Ellis) Barr, comb. nov. FIG. 83

Gnomonia magnoliae Ellis, Amer. Nat. **17**: 318. 1883.

Perithecia 220-440 μm diam, 110-200 μm high, in dull yellowish areas of leaf, wall reddish brown above, beaks erumpent through either leaf surface, 110-300 μm long, 40-70 μm wide near base, yellowish to reddish brown. Asci 36-52 × 5-7.5 μm. Ascospores (15.5-)18-26(-32) × (1-)2-2.5 (-3) μm, appendages 2-3.5 × 1 μm.

In leaves of *Magnolia virginiana* L., North America.

Material examined: *New Jersey:* Newfield, Jul 1882, J.B. Ellis (type, NY, BPI); Ell. & Ev. N.A.F. 1354 (BPI, MASS, MICH); *Maryland:* Suitland, 2 Jun 1928, W.W. Diehl (BPI); Herald Harbor, Anne Arundel Co., 1 Jul 1928, J.A. Stevenson & W.W. Diehl (BPI); *Georgia:* GA 6603, 6604, 6605 (GA).

23. **Pleuroceras** Riess, Hedwigia **1**: 25. 1854.

Cryptoderis Auersw., Mycol. Europ. **5/6**: 29. 1869.

Stromatic tissues not formed; perithecia immersed, oblique or horizontal in substrate tissues, beaks lateral, short to elongate, narrow or broad. Asci elongate to cylindroid. Ascospores hyaline, elongate, narrowed slightly toward the median septum and often separating readily at septum, with the part-spores one celled or one septate ("doubly fusoid"), or ±cylindroid, or tapered from wider upper end to base, appendages elongate and narrow when formed, in a fascicle in the ascus.

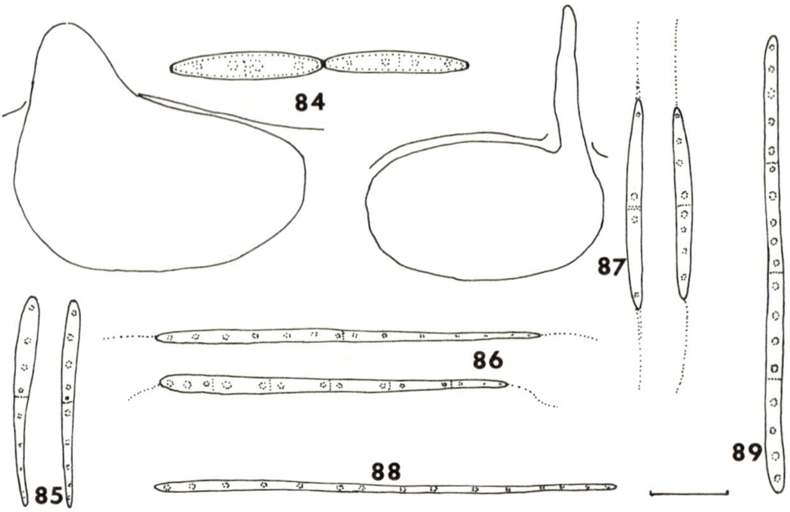

FIGS. 84-89. FIGS. 84-88. Species of *Pleuroceras*. 84. *P. cryptoderis*, outline of perithecium, ascospore. 85. *P. lirellaeformis*, ascospores. 86. *P. pleurostyla*, ascospores. 87. *P. tenella*, outline of perithecium, ascospores. 88. *P. sassafras*, ascospore. FIG. 89. *Plagiosphaera umbelliferarum*, ascospore. Scale =100 μm or 10 μm.

Conidial state: *Asteroma* (as *Septomyxa, Cylindrosporella*) where known.

In overwintered leaves of deciduous trees.

Type species: *Pleuroceras cryptoderis* (Lev.) v. Höhnel.

Cryptoderis is an obligate synonym of *Pleuroceras*. Lindau (**1897**) synonymized *Cryptoderis* with *Ophiognomonia*, but the two genera differ especially in position of perithecia and insertion of beak. Von Höhnel (**1918**) included *Linospora* as a synonym of *Pleuroceras* because he accepted as type species *L. candida* Fckl. However, *Linospora* is typified by *L. capreae* (see discussion under *Linospora*).

The "doubly fusoid" ascospores of the type species do not occur in all species in the present concept of the genus. In the following key to species the major diagnostic characters are utilized.

Key to Species of *Pleuroceras*

1. Ascospores narrowly "doubly fusoid," tapered to median septum and to ends; beaks narrow; on *Populus* 2
1. Ascospores narrowly cylindroid or broader above and tapered toward base 3
   2. Ascospores 37-40 × 2.5-3 μm, one(three)septate.......... 2. *P. cryptoderis*
   2. Ascospores 48-70 × 3.5-4 μm, three septate .... *P. populi*
3. Ascospores broadest in upper third, tapering toward base; beaks broad 4
3. Ascospores ±cylindroid throughout; beaks narrow or broad 6
   4. Ascospores 22-32 × 1-2 μm, one septate; on *Quercus* .... 2. *P. lirellaeformis*
   4. Ascospores longer; on *Salix* 5
5. Ascospores 48-52 × 4-5 μm, two septate .... 3. *P. groenlandica*
5. Ascospores 48-63(-72) × 1.5-2.5 μm, three septate.......... 4. *P. pleurostyla*
   6. Beaks broad; ascospores 135-180 × 1 μm; on *Salix* .... 5. *P. helvetica*
   6. Beaks narrow; ascospores shorter 7
7. Ascospores 20-36 × 1-1.5(-2) μm; on *Acer* .... 6. *P. tenella*
7. Ascospores 36-65 × 1 μm; on *Sassafras* .... 7. *P. sassafras*

1. **Pleuroceras cryptoderis** (Lev.) v. Höhnel, Ann. Mycol. **16**: 101. 1918. FIG. 84

Sphaeria cryptoderis Lev., Ann. Sci. Nat. Bot. ser. 3, **9**: 144. 1848.

Sphaeria lamprotheca Desm., Ann. Sci. Nat. Bot. ser. 3, **16**: 315. 1851.

Cryptoderis lamprotheca (Desm.) Auersw., Mycol. Europ. **5/6**: 29. 1869.

Linospora candida Fuckel, Symb. Mycol. p. 123. 1870.

Pleuroceras ciliatum Riess, Hedwigia **1**: 25. 1854.

Perithecia (275-)330-440 μm wide, 165-220(-275) μm high, beaks (40-) 100-330 μm long, 40-52 μm wide near base, leaf epidermis often reddened above perithecia. Asci ca. 80-90 × 10-12 μm. Ascospores 36-46 × 2-3 μm, elongate, tapering to blunt truncate ends and median septum, separating readily at septum, each part-cell one celled or one septate, ends and separation area thickened and refractive, elongate narrow appendages visible in some collections.

In overwintered leaves of *Populus alba* L., Europe.

Material examined: *Europe:* Riess 1822 (BPI); Linhart, Fungi hungarici 369 (BPI); Herb. Barbey-Boissier 163 (BPI); Herb. Boissier 731 (BPI).

Pleuroceras populi Thompson (Mycologia **46**: 655. 1954) occurs on leaves of *Populus tremuloides* Michx. in New York state (type specimen, Ithaca, 25 May 1937, GA), and is similar to *P. cryptoderis* except for larger asci (144-217 × 16-19 μm) and ascospores (48-70 × 3.5-4 μm) which are septate in each part-spore and have a long narrow appendage at each end. The conidial state has been found on *P. tremuloides* and *P. grandidentata* Michx. from Ontario, New York, New Jersey, Pennsylvania, and Wisconsin; this is *Septomyxa populina* (Peck) v. Arx, with acervuli often forming concentric rings in living leaves, conidia 22.5-35 × 2.5-3 μm, hyaline, cylindroid, one septate. Thompson (**1954**) obtained the conidial state in culture from ascospores.

2. **Pleuroceras lirellaeformis** (Pass.) Barr, comb. nov. FIG. 85

Gnomònia lirellaeformis Pass., Rev. Mycol. **8**: 33. 1880.

Perithecia 140-250(-400) μm diam, depressed globose, beaks 78-440 μm long, 40-80 μm wide near base, erumpent mostly through lower leaf epidermis. Asci 38.5-46 × 5-6.5 μm. Ascospores 22-32 × 1-2 μm, broadest above, tapering to pointed base, septum ±median.

In overwintered leaves of *Quercus* spp., Europe, North America.

Material examined: *North America: New Hampshire:* Barr 3819a, 3888; *Massachusetts:* Barr 6070; *California:* Barr 5912b (all MASS).

Passerini's original description fits the North American collections well. Saccardo (**1882**) queried "An *Hypospila*?," but the narrowly clavate ascospores with nearly median septum differ from those of *Hypospilina*.

3. **Pleuroceras groenlandica** (Rostr.) Barr, comb. nov.

Hypospila groenlandica Rostrup, Meddel. om Grønland **3**: 561. 1888.

Perithecia large, beaks short, lateral. Asci 95-115 × 10-12 μm. Ascospores 48-52 × 4-5 μm, cylindroid with upper end enlarged, tapering to pointed base, two septate.

In overwintered leaves of *Salix* spp., subarctic Greenland, Iceland, Sweden.

Material examined: *Europe:* Vestergren, Microm. rar. sel. 1223 (GA).

Nannfeldt (**1931**) pointed out that the lack of a clypeus in this species necessitated its removal from *Hypospila*. He also noted that the conidial state was *Cylindrosporella vleugeliana* (Bub.) Nannf. (*Gloeosporium vleugeliana* Bub. Svensk Bot. Tids. **5**: 345. 1911), with subcuticular acervuli and conidia 9-16 × 2.5-3.5 μm.

4. **Pleuroceras pleurostyla** (Auersw.) Barr, comb. nov. FIG. 86

Gnomonia pleurostyla Auersw., Mycol. Europ. **5/6**: 28. 1869.

Cryptoderis pleurostyla (Auersw.) Winter in Rabenh., Kryptogamenfl. **1**(2): 591. 1887.

Perithecia 300-440 μm diam, 275-330 μm high, horizontal, beaks lateral, 215-770 μm long, 52-96 μm wide near base, tapering to 37-56 μm wide near apex. Asci 67-90(-120) × 5-9 μm. Ascospores (35-)40-63(-72) × 1-2.5 μm, cylindroid, broader in the upper third, 1.5-2 μm, tapering to pointed base, 0.5-1 μm wide below, one, three, or five septate, delicate narrow appendages 10-12 μm long at upper end, 5-6 μm long at basal end, parallel in the ascus.

In overwintered leaves of *Salix* spp., Europe, North America.

Material examined: *Europe:* Krieger, Fungi sax. 2370 (TRTC). *North America: Newfoundland:* Hebron vicinity, Labrador, 24 Jul 1954, R.T. Wilce (MASS); *Oregon:* Barr 6284 (MASS).

5. **Pleuroceras helvetica** (Rehm) Barr, comb. nov.

Ophiognomonia helvetica Rehm, Ann. Mycol. **5**: 543. 1907.

Perithecia 500-560 μm diam, beaks broad, 140-280 μm long, 105-150 μm wide, tapering to 75 μm near apex, reddish brown. Asci 180-245 × 6-9 μm. Ascospores 135-180 × 1 μm, cylindroid, multiguttulate, no distinct septa seen, parallel in the ascus.

In dead leaves of *Salix* spp., Europe, North America.

Material examined: *North America: Quebec:* McLellan Strait, 20 Jul 1955, R.T. Wilce (MASS).

The collection is in accordance with the description of the species from Graubünden. It appears to be a subarctic-subalpine species.

6. **Pleuroceras tenella** (Ell. & Ev.) Barr, comb. nov. FIG. 87

Gnomonia tenella Ell. & Ev., J. Mycol. **4**: 80. 1888.

Perithecia 220-350 μm diam, 150-250 μm high, beak 280-385 μm long, 26-40 μm wide near base. Asci 60-110 × 4-7 μm. Ascospores 20-36 × 1-2 μm, cylindroid, septum median, appendages 10-30 × 1 μm, in a fascicle in the ascus.

In leaf blades, veins, and upper portion of petioles of *Acer rubrum* L., at times in other species of *Acer*, North America.

Material examined: *Ontario:* Malloch 647 (MASS); *Maine:* Barr 3261, 3420a (MASS); *New Hampshire:* Barr 3816b, 5159 (MASS); *Massachusetts:* Barr 5294, 5305, 5596, 5605, 6055 (MASS); *New York:* Barr 1888b (MASS); Cain 647 (TRTC); *New Jersey:* Newfield, Jun 1888, J.B. Ellis (type, NY); May 1890 (NY); Ell. & Ev. N.A.F. 2140 (BPI, DAOM, MASS, MICH); *Michigan:* Barr 886 (MASS); *Georgia:* GA 6695, 6696, 6697, 6698, 6699, 6700 (GA); *Tennessee:* GA 6702 (GA); *Louisiana:* Barr 6339 (MASS).

Pleuroceras tenella often occurs in the same leaves as *Plagiostoma acerophila.* The range of the latter species rarely extends as far south as that of *Pleuroceras tenella.*

7. **Pleuroceras sassafras** (Ell. & Ev.) Barr, comb. nov. FIG. 88

Gnomonia sassafras Ell. & Ev., Bull. Torrey Bot. Club **10**: 98. 1883.

Perithecia 220-300 μm diam, 170-200 μm high, beaks lateral, elongate, 450-800 μm long, 30 μm wide near base. Asci 50-70 × 3-5.5 μm. Ascospores 38-65 × 1 μm, cylindroid or slightly enlarged above, containing numerous guttules, no septum seen.

In leaves of *Sassafras albidum* (Nutt.) Nees var. *molle* (Raf.) Fern., North America.

Material examined: *New Jersey:* Ell. & Ev. N.A.F. 1684a (NY); *Ohio:* Fairfield Co., May 1883, W.A. Kellerman (type, NY); Ell & Ev. N.A.F. 1684 (MICH, NY).

24. **Plagiosphaera** Petrak, Ann. Mycol. **39**: 289. 1941.

Perithecia immersed, oblique or horizontal, thickly scattered, beaks lateral, often curved, relatively short and broad; peridium narrow. Asci cylindroid clavoid. Ascospores hyaline, elongate cylindroid, indistinctly septate, not appendaged, in fascicle in the ascus.

In herbaceous stalks.

Type species: *Plagiosphaera immersa* (Trail) Petrak.

Petrak separated *Plagiosphaera* from *Ophiognomonia* by position of perithecia in the substrate and insertion of beaks. *Plagiosphaera* is more closely related to *Pleuroceras* by these features than to *Ophiognomonia.* The habit of species of the genera is one separating character: in herbaceous stalks for *Plagiosphaera,* in overwintered leaves of deciduous trees for *Pleuroceras* (and *Ophiognomonia*). The perithecial beaks of *Pleuroceras* are narrower and usually more elongate than those of *Plagiosphaera.* Ascospores of *Plagiosphaera* species are cylindroid with rounded ends, usually curved, and delicately septate, like those of *Sillia.*

1. **Plagiosphaera immersa** (Trail) Petrak, Sydowia **14**: 351. 1960.

Ophiobolus immersus Trail, Trans. Bot. Soc. Edinb. **17**: 492. 1889.

Ophiobolus brachysporus Fautrey & Roum., Rev. Mycol. **14**: 109. 1892.

Ophiobolus moravicus Petrak, Ann. Mycol. **19**: 80. 1921.

Plagiosphaera moravica (Petrak) Petrak, Ann. Mycol. **39**: 289. 1941.

Perithecia elliptic, 300-600 μm long, 100-200 μm wide, beaks relatively short, 100-150 × 70-80 μm. Asci 60-100 × 7-10 μm. Ascospores (40-) 50-65 × 2-2.5(-3) μm, cylindroid, guttulate, finally delicately septate, up to six septate, in parallel fascicle in the ascus.

In overwintered stalks of *Campanula* sp., *Urtica* sp., Europe.

Dennis (**1975**) redescribed and illustrated this species from Great Britain. It has not yet been collected in North America.

Linocarpon umbelliferarum Barr must be transferred to *Plagiosphaera* as **Plagiosphaera umbelliferarum** (Barr) Barr, comb. nov. (Basionym: *Linocarpon umbelliferarum* Barr, Canad. J. Bot. **39**: 320. 1961) (FIG. 89). The perithecia are oblique and the short broad beaks eccentric in the substrate; no clypeus typical of *Linocarpon* is formed. A description and illustration were provided by Barr (**1961**). *Gaeumanno-*

myces leptosporus Iqbal (Trans. Brit. Mycol. Soc. **58**: 346. 1972), described from Great Britain in umbellifer stalks immersed in water, is closely related to *P. umbelliferarum*. The perithecia in culture were described as oblique and the beaks lateral (Iqbal, **1972**). The ascospores are longer (60-90 μm) than those of *P. umbelliferarum* (52-67 μm), and develop more septa, but there is scarcely a specific difference between the two.

II-2. **Valsaceae** subfamily **Valsoideae**

Perithecia single or grouped in substrate, beaks erumpent separately or converging, stromatic tissues prosenchymatous, as blackened zone or clypeus or stromatic disc at substrate surface, or also with blackened marginal zone in substrate. (3 tribes).

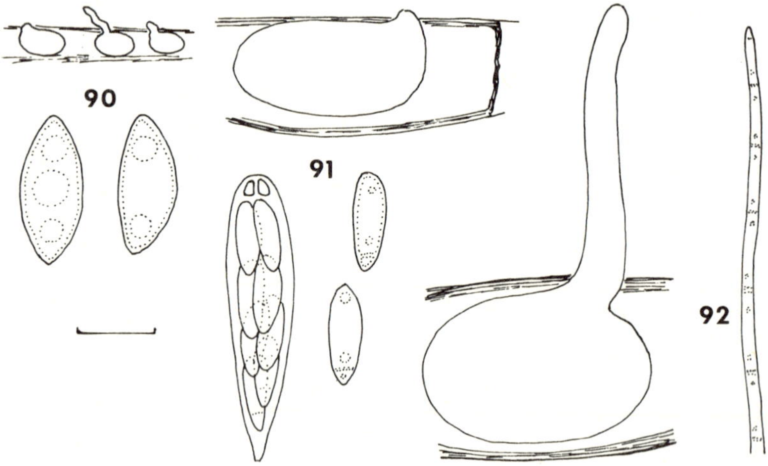

FIGS. 90-92. FIG. 90. *Clypeoporthella kriegeriana,* perithecia in section, ascospores. FIG. 91. *Hypospilina bifrons,* outline of perithecium, ascus, ascospores. FIG. 92. *Linospora capreae,* outline of perithecium, portion of ascospore. Scale =1 mm, 100 μm, or 10 μm.

II-2a. Tribe **Linosporeae** Barr, tribus nov.

Perithecia singulara vel pauciaggregata, stromata instar zonae nigrifactae vel clypei; in foliis et petiolis dicotyledonibus arboribus et frutibus vel herbarum caulis vel monocotyledonibus. Genus typicus: *Linospora* Fckl.

Key to Genera of Tribe Linosporeae

1. Ascospores one celled, ellipsoid or fusoid .... 25. *Clypeoporthella*
1. Ascospores septate or elongate if not septate 2
    2. Ascospores ellipsoid or ovoid, apiosporous.................... 26. *Hypospilina*
    2. Ascospores filiform or cylindroid or acicular, nonseptate or one to several septate... .... 27. *Linospora*

Stioclettia Dennis (**1975**) would appear to belong in this tribe. In *S. luzulina* Dennis, the type species, the gregarious perithecia are oblique in hyphal stromatic tissues and have a short papillate lateral beak. Ascospores are oblong fusoid and become five septate.

25. **Clypeoporthella** Petrak, Ann. Mycol. **22**: 149. 1924.

Batschiella Kirschstein, Kryptogamenfl. Mark Brandenb. 7(3): 324. 1938.

Stromatic tissues prosenchymatous, sparse and mixed with cortical cells of substrate, often forming a blackened upper marginal zone in epidermal cells; perithecia oblique or horizontal, usually separate, in small groups at times, immersed, beaks lateral or eccentric, relatively short and broad. Asci ellipsoid. Ascospores hyaline, faintly pigmented in age, ellipsoid or fusoid, one celled, contents with one, three, or several globules, wall smooth.

Conidial state: *Phomopsis*-like.

In overwintered herbaceous stalks.

Type species: *Clypeoporthella brencklei* Petrak.

The type species of *Batschiella*, *B. lysimachiae* Kirschst., is identical with *Phomatospora kriegeriana* Rehm, which Petrak (**1940**) transferred to *Clypeoporthella*. This species had been placed in *Diaporthopsis* by von Höhnel (**1918**) and was retained there by von Arx and Müller (**1954**), although Wehmeyer (**1933**) considered it to be a doubtful species.

1. **Clypeoporthella brencklei** Petrak, Ann. Mycol. **22**: 149. 1924.

Diaporthopsis brencklei (Petrak) v. Arx & Müller, Kryptogamenfl. Schweiz **11**(1): 372. 1954.

Perithecia 300-400 μm diam. Asci 50-75 × 8-11 μm. Ascospores 10-18 × 4-6 μ.

In stalks of *Aster* spp., North America.

I have been unable to locate collections of this species.

2. **Clypeoporthella kriegeriana** (Rehm) Petrak, Ann. Mycol. **38**: 199. 1940. FIG. 90

Phomatospora kriegeriana Rehm, Ann. Mycol. **10**: 390. 1912.

Diaporthopsis kriegeriana (Rehm) v. Höhnel, Ann. Mycol. **16**: 113. 1918.

Batschiella lysimachiae Kirschst., Kryptogamenfl. Mark Brandenb. 7(3): 324. 1938.

Perithecia ±500 μm diam. Asci 76-90 × 12-16 μm. Ascospores 14-21 × 7-9 μm.

In overwintered stalks of *Lysimachia* spp., Europe, North America.

Material examined: *North America: Quebec:* Barr 2161 (MASS).

Although von Arx and Müller (**1954**) included *Clypeoporthella* in the synonymy of *Diaporthopsis*, the perithecia of species in the latter genus are upright and beaks are central. *Clypeoporthella* is retained for species whose perithecia are oblique to horizontal, separate or few grouped beneath blackened substrate surface. *Diaporthopsis appendiculata* Wehm. (The Genus *Diaporthe*, p. 233. 1933) on *Aster* sp., has ascospores larger than those of either *C. brencklei* or *C. kriegeriana*, 24-28 × 5.5-7 μm, with an elongate narrow appendage at each end, and appears to belong in *Clypeoporthella* rather than *Diaporthopsis* because of the oblique perithecia and lateral beaks.

26. **Hypospilina** (Sacc.) Traverso, Flora Ital. Crypt. Pars **I**: 495. 1913.

Hypospila subgenus *Hypospilina* Sacc., Syll. Fung. **2**: 190. 1883.

Hypospila auct. non Fries, Syst. Orbis Veget. p. 109. 1825.

Stromatic tissues prosenchymatous, forming small clypei in both epidermal layers, connected by lateral marginal zones, these composed of compact dark hyphae in leaf cells; perithecia one or few beneath clypeus, oblique or lateral, beaks short, eccentric or lateral. Asci ellipsoid. Ascospores hyaline, ellipsoid, apiosporous.

In dead leaves of frondose trees.

Type species: *Hypospilina bifrons* (DC. ex Fries) Trav.

Hypospila Fries was based upon *Spiloma inustum* Ach. (*Sphaeria inusta* Fries, Elenchus Fung. **2**: 88. 1828). Fries noted that the species, collected on branches of a tropical tree, was related to *Dichaena*, and was apparently lichenized. However, in **1849** Fries added to the genus *H. quercina* or *bifrons* and *H. populina* or *ceuthocarpa*, thus creating a mixed concept of *Hypospila*. Karsten (**1873**) added *H. pustula* and Saccardo (**1883**) recognized two subgenera: *Hypospila*, with ascospores becoming three septate, for *H. pustula* (Pers.) Karst. and *H. rehmii* Sacc.; *Hypospilina* with ascospores one septate for *H. bifrons* and *H. immunda*. Traverso (**1913**) raised *Hypospilina* to generic level, and typified it by *H. bifrons*. Von Höhnel (**1918**) suggested retaining *Hypospila* with the new typification of *H. bifrons*, as did von Arx (**1951**) and Schrantz (**1960**). Petrak and Sydow (**1923**) reviewed the complicated nomenclatural situation of *Hypospila*. They studied original material of *Spiloma inustum* and found it to consist of sterile spot-like "stromata" on whitish areas of bark. They recommended that *Hypospila* Fries be stricken and that *Hypospilina* (Sacc.) Trav. be utilized for *H. bifrons*. Clements and Shear (**1931**) typified *Hypospila* by *H. pustula*, but von Arx (**1951**) showed that that species (as well as *H. rehmii*) belongs in *Plagiostoma*. Finally, Müller and von Arx (**1962**, **1973**), following Petrak and Sydow (**1923**) used *Hypospilina* with *H. bifrons*.

Hypospila immunda (Fckl.) Sacc., the second species in Saccardo's subgenus *Hypospilina* (**1883**), also occurs on leaves of *Quercus*. Von Höhnel (**1918**) transferred the species to *Didymella*, noting that a clypeus was produced. From the description, in particular the information on numerous, filiform paraphyses, cylindric asci with an amyloid pore area, it appears to belong in *Paradidymella*, probably identical with *P. succinea* (Rob.) Petrak.

1. **Hypospilina bifrons** (DC. ex Fries) Trav., Flora Ital. Crypt. Pars **1**: 495. 1913. FIG. 91

[*Xyloma bifrons* DC., Fl. Franc. **6**: 156. 1815.]

Sphaeria (Depazea) bifrons DC. ex Fries, Syst. Mycol. **2**: 438. 1823.

Hypospila quercina bifrons (DC. ex Fries) Fries, Summa Veget. Scand. p. 421. 1849.

Gnomonia bifrons (DC. ex Fries) Auersw., Mycol. Europ. **5/6**: 20. 1869.

Hypospila bifrons (DC. ex Fries) Sacc., Syll. Fung. **2**: 190. 1883.

Stromatic areas black on both leaf surfaces, angular, 0.5-1 mm wide, containing 1-4(-7) perithecia, 156-275 μm diam, 104-165 μm high, beak short, lateral, erumpent epiphyllously. Asci 35-40 × 8-10 μm, apical annulus 2-3 μm wide and high. Ascospores 10-14 × 3.5-5 μm, hyaline, ellipsoid or ovoid with tapered fusoid ends, straight or inequilateral, septum in lower third of spore, lower cell 1.5-3 μm long and wide, overlapping biseriate in the ascus.

In overwintered leaves of *Quercus* spp., Europe. Ellis and Everhart (**1892**) cited a record by Schweinitz from Bethlehem, Pennsylvania, but I have not seen any collections of this fungus from North America.

Material examined: *Europe:* Saccardo, Mycotheca ital. 649; Deutschlands Schwämme, Holl, Schmidt & Kunze CCIV; Allescher & Schnabl, Fungi bavarici 443; Vestergren, Micromycetes rar. sel. 916 (*Gnomonia setacea* also present); Herbier Barbey-Boissier 149 (all BPI).

Hypospilina salicina Müller (Sydowia **11**: 468. 1957), described with large, one-celled ascospores, deviates from the generic concept, so that it should be transferred to another genus, perhaps *Clypeoporthella.*

27. **Linospora** Fuckel, Symb. Mycol. p. 123. 1870.

Ceuthocarpon Karsten, Mycol. Fenn. **2**: 22. 1873.

Stromatic tissues prosenchymatous, in leaf interior composed of hyphae mixed with substrate cells, compact in epidermal cells and forming a blackened clypeus, often lateral marginal clypei also; perithecia immersed, oblique or horizontal, beak lateral, short to elongate. Asci elongate cylindroid. Ascospores filiform or cylindroid or tapering to base, hyaline, nonseptate or one to several septate.

Conidial state: *Melasmia*-like where known.

In overwintered leaves of deciduous trees.

Lectotype species: *Linospora capreae* (DC. ex Fries) Fuckel.

Fuckel (**1870**) included the species *L. candida, L. capreae, L. vulgaris, L. tigrina, L. procumbens* in *Linospora*, without specifiying a type species, but in the generic description he observed that the sterile stroma was " Phomam salicinam Depazeamque Fr.". *Sphaeria salicina* Sowerby was one of the names Fries (**1823**) cited under *Phoma saligna* Ehrh. ex Fries, and Fuckel specified that *Phoma saligna* was the sterile stroma of *L. capreae*. Thus, the typification of *Linospora* by *L. capreae* would meet Fuckel's concept of the genus. *Linospora candida* on *Populus* is a synonym of *Pleuroceras cryptoderis;* the other three species were all described from *Salix* leaves. Traverso (**1906**) reasoned that *Sphaeria saligna* Ehrh. was earlier than *S. capreae* DC., and proposed the name *Linospora saligna* (Ehrh. ex Fries) Trav. to replace *L. capreae* (DC. ex Fries) Fuckel. However, *Sphaeria capreae* DC. ex Fries is the earliest legitimate epithet for the species, since Fries specified in his generic diagnosis of *Phoma* "Asci nulli." Von Höhnel (**1917**) utilized *Phoma* Fries with *P. capreae* (DC.) v. Höhnel specified as the type species, and reduced *Linospora*, which he typified by *L. candidum*, to synonymy with *Pleuroceras. Phoma* Desm. (1849) ss. Saccardo et auct. is recommended for conservation against *Phoma* Fries (**1823**) (Trans. Brit. Mycol. Soc. **23**: 289. 1939.)

Ceuthocarpon, with *C. populina* (Fr.) Karsten, was erected for *Linospora*-like species which lacked an elongate beak. Winter (**1887**) and Lindau (**1897**) recognized that this character is a variable one and placed *Ceuthocarpon populinum* in *Linospora*. Von Höhnel (**1917**) retained Karsten's genus and (**1918**) noted that the stromatic tissues were carbonaceous and the beak very short and broad, to separate the genus from *Linospora* (*Phoma* in his sense). Von Arx (**1951**) preferred to recognize only *Linospora*, and synonymized *Ceuthocarpon* and *Linospora*. According to Sydow's Mycotheca germ. 246, *L. populina* is indeed a species of *Linospora*, *L. ceuthocarpa* (Fr.) Lind. (The combination made by Morelet in Bull. Soc. Sci. Nat. Archeol. Toulon Var **202**: 11. 1973, was unnecessary.)

1. **Linospora capreae** (DC. ex Fries) Fuckel, Symb. Mycol. p. 124. 1870.
FIG. 92

Sphaeria capreae DC. ex Fries, Syst. Mycol. **2**: 517. 1823.

Linospora saligna (Ehrh. ex Fries) Trav., Flora Ital. Crypt. Pars **I**: 350. 1906.

Phoma capreae (DC. ex Fries) v. Höhnel, Ann. Mycol. **16**: 103. 1918.

Perithecia single or 2 to 3 beneath blackened clypeus, visible on both leaf surfaces, 260-400(-600) μ diam, 150-250 μm high, beaks erumpent through either leaf surface, at times two and through both, short to elongate, 200-1000 μm long, ca. 60 μm wide near base. Asci 130-300 × 6-10 μm. Ascospores 120-225 × 1-2(-3) μm, hyaline, filiform, up to 16 septate, parallel or often coiled in mid part of ascus.

In overwintered leaves of *Salix* spp., Europe, North America.

Material examined: *Europe:* Petrak, Flora Boh. et Morav. exs. 735 (GA). *North America: British Columbia:* Barr 27 (UBC).

Klebahn (**1918**) germinated ascospores which develop into slow-growing colonies, but neither conidia nor perithecia were produced. His infection studies, using fresh ascospores sprayed on leaves, gave only small yellow spots and no further development of the fungus, nor production of perithecia after the sprayed leaves were overwintered.

Linospora insularis Johans., Öfversigt Kongl. Vetensk.-Akad. Förh. **9**: 171. 1884 (*Ceuthocarpon insularis* (Johans.) Berl., Icones Fung. **2**: 149. 1909), has shorter ascospores, 75-112 × (1.5-)2.5-3 μm, with a single supramedian septum. It was described from Iceland and is known in North America from the Northwest Territories: Dansereau 5008101459b; Schuster 35279b (DAOM, MASS).

Linospora tetraspora G.E. Thompson, Canad. J. Res. C, **17**: 236. 1939, on *Populus* spp., has four spored asci, ascospores 155-200 × 1-2 μm, five to eight septate. The species is known from Quebec, Ontario, Alberta, British Columbia, and Vermont. The type collection is from Ontario: Sandy Inlet, Lake Timagami, 12 Jul 1931, G.E. Thompson, and is distributed in a number of herbaria (DAOM, GA, TRTC).

Linospora gleditschiae Miller & Wolf, Mycologia **28**: 177-179. 1936, occurs in overwintered leaves of *Gleditschia triacanthos* L., following the formation of acervuli of *Melasmia gleditschiae* (Lev.) Ell. & Ev. in living leaves (Miller and Wolf, **1936**). The clypeus is not particularly obvious in herbarium material, and the horizontal perithecia are reddish brown with relatively short beaks. Ascospores are nonseptate, 70-90 × 3 μm. Numbers GA 7005-7101 include specimens from Georgia and Oklahoma.

Two species ascribed to *Linospora* belong in the Loculoascomycetidae: *L. conflicta* (Cooke) Sacc., Syll. Fung. **2**: 355. 1883, is *Sphaerulina conflicta* (Cooke) Barr, Contrib. Univ. Mich. Herb. **9**: 607. 1972. *Linospora brunnellae* Ell. & Ev., Proc. Acad. Nat. Sci. Phila. for 1894, **46**: 337. 1895, was also described on p. 338 as *Hypospila brunnellae* Ell. & Ev. (*Phoma brunnellae* (Ell. & Ev.) Kuntze, Rev. Gen. Pl. **3**: 502. 1898; *Ceuthocarpon brunnellae* (Ell. & Ev.) Berl. Icones Fung. **2**: 149.

1899) is instead *Leptosphaeria hesperis* Barr, Canad. J. Bot. **45**: 1867, or *Ophiobolus prunellae* (Ell. & Ev.) Shoemaker, Canad. J. Bot. **54**: 2388. 1976.

Linospora ferruginea Ellis & Martin, Amer. Nat. **18**: 69. 1884 (*Ceuthocarpon ferrugineum* (Ell. & Mart.) Berl., Icon. Fung. **2**: 147, 153. 1899) must also be removed from *Linospora.* According to the holotype collection (NY, in leaves of *Andromeda ferruginea* Walt., Green Cove Springs, Florida, Dec 1882, Dr. Martin; also issued as Ellis N.A.F. 1300 (MASS, MICH, NY); Rehm Ascom. 942 (NY); Rabenh.-Wint. Fungi eur. 3456 (NY); Hume's Fungi of Florida 46 (NY); this species is related to *Ophiodothella leucospila* (Berk. & Curt.) Miller & Thompson, and the transfer to **Ophiodothella ferruginea** (Ell. & Mart.) Barr, comb. nov., is proposed. The upright, thin-walled perithecia develop singly or two or three in small pallid leaf spots; each beak is surrounded by a narrow blackened clypeus. At times the perithecial base, too, is covered by a narrow clypeus. Ascospores are narrow cylindroid, 33-54 × 1.5-2 (-3) μm; the septum is median when formed. *Ophiodothella leucospila* and *O. ferruginea* are part of a series of species in *Ophiodothella* where the stromatic clypeus is reduced in size, unlike that in the typical species of the genus.

II-2b. **Valsaceae** tribe **Valseae**

Perithecia grouped beneath stromatic disc, entostroma poorly or well developed; in branches of trees and shrubs. Type genus: *Valsa* Fries.

Key to Genera of Tribe Valseae

1. Ascospores narrow, cylindroid and usually allantoid, one celled or one to several septate 2
1. Ascospores ellipsoid or fusoid, not allantoid, one septate 3
   2. Ascospores one celled, usually not over 30 μm long .... 28. *Valsa*
   2. Ascospores one celled or one to several septate, usually longer than 30 μm .... 32. *Ophiovalsa*

3. Perithecia grouped beneath stromatic disc, occasionally separate, entostromatic tissues sparse; conidial locules usually formed in separate stromata from perithecia.. 29. *Cryptodiaporthe*

3. Perithecia grouped in abundant entostromatic tissues beneath stromatic disc; conidial locules formed in entostroma prior to perithecial formation...................... 4

4. Stromatic tissues dull colored, brown or grayish brown; ascospores narrowly ellipsoid fusoid.. 30. *Amphiporthe*

4. Stromatic tissues brightly colored, yellow, orange, or bright brownish; ascospores broadly ellipsoid.. 31. *Cryphonectria*

Apioporthella Petrak (Ann. Mycol. **27**: 401. 1929) with the type species *A. bavarica* Petrak, is known only from Europe at present. It is the apiosporous representative of the tribe Valseae. I have considered including *Cryptodiaporthe vepris* and related species in *Apioporthella* but the ascospores of these species, while unequally septate, are not truly apiosporous.

28. **Valsa** Fries, Summa Veget. Scand. p. 410, 1849, emend. Saccardo, Atti Soc. Ven. Trent. in Padova **4**: 4. 1875; Syll. Fung. **1**: 108. 1882.

Stromatic tissues prosenchymatous, entostroma sparsely or well developed, ectostromatic disc well or slightly developed; perithecia circinate to crowded, oblique or horizontal, beaks oblique or lateral, converging and erumpent through ectostromatic disc. Asci ellipsoid. Ascospores hyaline, allantoid or nearly cylindroid, one celled.

Conidial state: *Cytospora* Ehrenb. ex Fr. (*Cytophoma* v. Höhnel) spp.: conidial locules formed in stromata; conidia hyaline, allantoid or cylindroid, one celled.

In branches of deciduous tree and shrubs, and also coniferous trees.

Lectotype: *Valsa ambiens* (Pers. ex Fries) Fries.

Fries's (**1849**) concept of *Valsa* encompassed several disparate elements, not only diaporthaceous genera which are now segregated as separate genera (e.g. *Leucostoma, Allantoporthe*), but also valsoid species belonging in the Diatrypaceae (*Eutypella*) and in the Calosphaeriaceae (*Calosphaeria*). Nitschke (**1867**) had a broad concept of *Valsa* also, and Winter (**1887**) and Lindau (**1897**) retained and enlarged the concept. Saccardo (**1875, 1882**) emended and restricted the genus to the subgenera *Euvalsa* and *Leucostoma*. In *Euvalsa*, he utilized ascospore sizes to separate Microsporae and Macrosporae. These groupings are comparable to Nitschke's (**1867**) subgenera Monostichae and Cir-

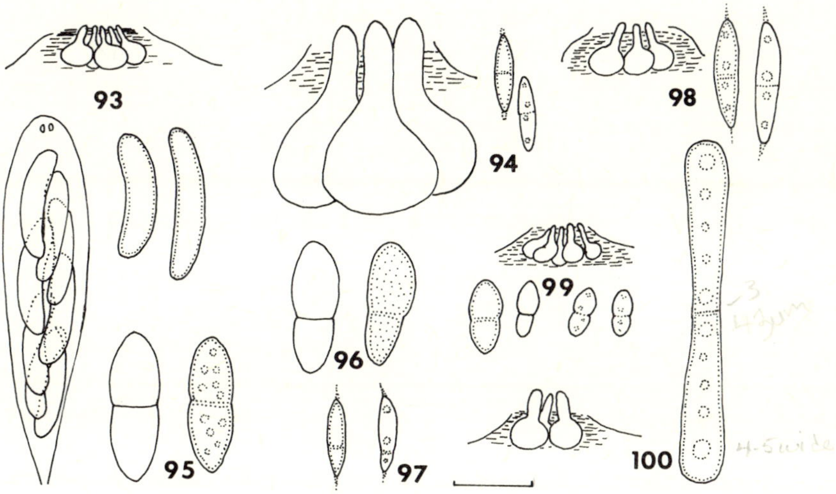

FIGS. 93-100. FIG. 93. *Valsa ambiens,* stroma with perithecia in section, ascus, ascospores. FIGS. 94-97. Species of *Cryptodiaporthe.* 94. *C. petiolophila,* stroma with perithecia in section, ascospores. 95. *C. salicella,* ascospores. 96. *C. corni,* ascospores. 97. *C. vepris,* ascospores. FIG. 98. *Amphiporthe aculeans,* stroma with perithecia in section, ascospores. FIG. 99. *Cryphonectria parasitica,* stroma with perithecia in section, ascospores. FIG. 100. *Ophiovalsa femoralis,* stroma with perithecia in section, ascospore. Scale =1 mm, 100 μm, or 10 μm.

cinatae, and are useful to separate the two groups of species now retained in *Valsa.* Von Höhnel (**1917**) designated *Valsa ambiens* (Pers. ex Fries) Fries, a representative of the Macrosporae or Circinatae, as type species of the genus.

Valsa section **Valsa**

Perithecia relatively few, large, circinate in sparse entostroma, ectostromatic disc well developed; ascospores mostly more than 10 μm long.

Type species: *Valsa ambiens* (Pers. ex Fries) Fries.

1. **Valsa ambiens** (Pers. ex Fries) Fries, Summa Veget. Scand. p. 412. 1849. FIG. 93

Stromatic disc gray, then blackened, stroma 0.5-3 mm diam; perithecia circinate, relatively few (up to 8), 280-500 μm diam, beaks converging. Asci 50-65(-80) × (6.5-)10-14.5 μm, eight spored or less than eight maturing. Ascospores (11.5-)16-25(-33) × (2.5-)4-6.5 μm.

Conidial state: Conidia 5-6.5(-9) × 1-2 μm.

In branches of deciduous trees, cosmopolitan.

Representative material examined: *North America: Massachusetts: Tilia* sp., Barr 3781 (MASS).

Valsa section **Monostichae** Nits.

Perithecia numerous, relatively small, crowded in well-developed entostromatic tissues, ectostromatic disc reduced; ascospores mostly less than 10 μm long.

Lectotype species: *Valsa ceratosperma* (Tode ex Fries) Maire.

2. **Valsa ceratosperma** (Tode ex Fries) Maire, Publ. Inst. Bot. 3: 20. 1937.

A long list of synonyms was provided by Défago (**1942**); Hubbes (**1960a**) and Kobayashi (**1970**) added a few more. An additional synonym is *Valsa rubi* Peck, New York State Museum Rep. for 1874, **28**: 72-73. "1875" 1876, non *Valsa rubi* (Fuckel, **1870**).

Stromata variable in size, 0.5-2 mm diam, brown, gray brown, reddish brown; perithecia relatively numerous, 165-360 μm diam, beaks elongate, converging. Asci 17-35 × 3-6 μm. Ascospores 4.5-8(-13) × 1-2(-3) μm.

Conidial stromata containing numerous locules; conidia 5-6.5 × 1 μm.

In branches of deciduous trees and shrubs, cosmopolitan.

Representative material examined: *North America: Quebec:* Barr 1992 (MASS); *New York:* Forestburgh, Sep, C.H. Peck (NYS, type of *Valsa rubi* Peck).

29. **Cryptodiaporthe** Petrak, Ann. Mycol. **19**: 118. 1921.

Chorostella (Sacc.) Clements & Shear, Genera of Fungi, p. 264. 1931.

Diaporthe subgenus *Chorostella* Sacc., Syll. Fung. **1**: 632. 1882.

Stromatic tissues prosenchymatous, forming disc through which converging beaks are erumpent, without or with slight development of entostromatic tissues; perithecia oblique, grouped and circinate in periderm, occasionally single and oblique or lateral, beaks eccentric or lateral, convergent. Asci ellipsoid. Ascospores hyaline, ellipsoid fusoid or broadly ellipsoid, septum median or submedian, appendages narrow, setose, when present.

Conidial state variable: borne in pycnidium-like locules in small or large stromata; conidiophores short, conidiogenous cells phialidic; conidia hyaline, rarely pigmented, nonseptate or one septate, ellipsoid or fusoid. Described under the generic names *Fusicoccum* Cda., *Discella* Berk. & Br. (=*Diplodina* Westd., Sutton, **1977**). *Chondroplea* Kleb. (=*Discosporium* v. Höhnel, Sutton, **1977**), *Zythia*, or *Neobarclayella* Sacc. (=*Polynema* Lév., Sutton, **1977**),

In woody branches, occassionally in petioles, of deciduous trees and shrubs.

Type species: *Cryptodiaporthe aesculi* (Fckl.) Petrak.

Chorostella was typified by *C. hystrix* (Tode ex Fries) Clements & Shear. This species, on *Acer* (and *Alnus*?) in Europe, has ascospores which are typical of *Cryptodiaporthe*. The perithecia suggest those of *Gnomonia*, according to Wehmeyer (**1933**), but there are some stromatic tissues formed above grouped perithecia.

Wehmeyer (**1933**) considered that *Cryptodiaporthe* contained a heterogenous group of species. Some of the species which he accepted have been or should be transferred to other genera; other species are added by recognition of the true nature of *Apioporthe* (see *Anisogramma*). The following is a suggestion toward clarification of the relationships of species of *Cryptodiaporthe* — separate genera may ultimately be needed to emphasize differences between these sections.

Cryptodiaporthe section **Cryptodiaporthe**

Ascospores ellipsoid fusoid, narrow, tapering to ±pointed ends, septum ±median; conidial state: *Fusicoccum* or *Discella* where known.

Type species: *Cryptodiaporthe aesculi* (Fckl.) Petrak.

1. **Cryptodiaporthe aesculi** (Fckl.) Petrak, Ann. Mycol. **19**: 119. 1921.

Ascospores 14-23 × 4.5-7 μm; in branches of *Aesculus hippocastanum* L., Europe.

2. **Cryptodiaporthe macounii** (Dearn.) Wehm., The Genus Diaporthe, p. 191. 1933.

Ascospores 7.5-10 × 1.5-2.5 μm; in branches of *Spiraea* spp., eastern and western North America. (Material examined: *New Hampshire:* Barr 3821, MASS).

3. **Cryptodiaporthe petiolophila** (Peck) Barr, comb. nov. FIG. 94

Sphaeria petiolophila Peck, New York State Museum Rep. **35**: 144, 145. (for 1881) 1884.

Gnomonia petiolophila (Peck) Berl. & Vogl., Syll. Fung. Addit. **1-4**: 90. 1886; Syll. Fung. **9**: 674. 1891.

Stromatic tissues prosenchymatous, forming a disc around beaks of separate or small groups of perithecia, hyphal around and beneath perithecia; perithecia 150-250(-300) μm diam, beak 117-156 μm long, 60-104 μm wide; peridium 11-16.5 μm wide, blackish brown externally, light brown internally. Asci 25-36 × 5.5-7 μm, apical annulus narrow. Ascospores 7-12 × 1-2.5 μm, narrowly ellipsoid or upper portion enlarged, straight or slightly curved, septum median, appendages 1.5-2 μm long, 1 μm wide, rounded.

In petioles and small twigs of *Acer saccharum* Marsh., *A. spicatum* Lam., occasionally in *Carya tomentosa* Nutt., eastern North America..

Material examined: (Representative collections): *Ontario:* TRTC 45712 (TRTC); Granton, 30 Apr 1894, Dearness (NY); *New York:* Adirondack Mts., Jun 1881, C.H. Peck (type); Karner, 11 Apr 1914, H.D. House; Oneida, 15 May 1918, H.D. House (all NYS).

Additional synonyms include *Diaporthe spicata* Ell. & Ev., *Cryptodiaporthe densissima* var. *spicata* (Ell. & Ev.) Wehm.

4. **Cryptodiaporthe apiculata** (Wallr.) Petrak, Ann. Mycol. **19**: 176. 1921.

Ascospores 15-23 × 2-3.5 μm; in branches of *Salix* and *Populus* spp., Europe, western North America. (*C. salicella* ss. Wehmeyer, **1933**). (Material examined: *Washington:* Northport, 23 May 1935, G.G. Hedgcock, BPI).

5. **Cryptodiaporthe galericulata** (Tul.) Wehm., The Genus Diaporthe, p. 211. 1933.

Ascospores 18-26 × 4-5.5 μm; in branches of *Fagus* spp., Europe, eastern North America.

Cryptodiaporthe section **Latispora** Barr, sect. nov.

Ascosporae ellipsoideae, latae, septo medio vel inframedio. Species typica: *Cryptodiaporthe salicella* (Fr.) Petrak.

Ascospores broadly ellipsoid, ends rounded, septum median or slightly submedian; conidial state *Discella*.

6. **Cryptodiaporthe salicella** (Fries) Petrak, Ann. Mycol. **19**: 180. 1921. FIG. 95

Ascospores (11-)15-20 × 4.5-6.5 μm; in branches of *Salix* spp., Europe, North America. (*C. salicina* ss. Wehmeyer, **1933**). (Material examined: *Quebec:* Barr 2107, MASS; *Massachusetts:* Barr 5148, MASS).

7. **Cryptodiaporthe acerinum** Reid & Cain, Canad. J. Bot. **40**: 839. 1962.

Ascospores 13-18 × 5-7 μm; in branches of *Acer spicatum* Lam., *A. rubrum* L., eastern North America. (Material examined: *Massachusetts:* Barr 5161, MASS; *New Jersey:* Ellis N.A.F. 180, as *Valsa myinda* Cke. & Ellis, MASS).

Cryptodiaporthe section **Noxa** Barr, sect. nov.

Ascosporae ellipsoideae, latae, septo medio vel inframedio; fungi parasitici. Species typica: *Cryptodiaporthe populea* (Sacc.) Butin.

Ascospores broadly ellipsoid, ends rounded, septum median or slightly submedian; conidial state *Chondroplea, Zythia.*

The two species noted below are parasitic; infection by the species causes death of its respective host tree.

8. **Cryptodiaporthe populea** (Sacc.) Butin, Nachtr. Bl. Deutsch. Pflanzenschutzd. (Braunschweig) **9**: 69. 1957.

Ascospores 14-20 × 6-8 μm; in branches of *Populus* spp., especially Lombardy poplar, Europe, North America. Conidial state: *Chondroplea populea* (Sacc.) Kleb. (Material examined: *Massachusetts:* Barr 6053, MASS).

9. **Cryptodiaporthe corni** (Wehm.) Petrak, Ann. Mycol. **39**: 302. 1941.
FIG. 96

Ascospores 12-17.5 × (2.5-)5-7 μm; in branches of *Cornus alternifolia* L.f., eastern North America. Conidial state: *Zythia aurantiaca* (Peck) Sacc. (Material examined: *Vermont:* Barr 4598, MASS; *Massachusetts:* Barr 2929, 5121, 6038, MASS).

Cryptodiaporthe section **Inaequalitas** Barr, sect. nov.

Ascosporae ellipsoideae vel fusoideae, comparate angustae, septo inframedio, obstipae. Species typica: *Cryptodiaporthe aubertii* (West.) Wehm.

Ascospores relatively narrowly ellipsoid or fusoid, bent at submedian septum; conidial state: *Phomopsis*-like or *Neobarclayella*-like where known.

10. **Cryptodiaporthe aubertii** (West.) Wehm., The Genus Diaporthe, p. 202. 1933, var. **aubertii.**

Ascospores 13-16.5(-20) × 2.5-5 μm; in branches of *Myrica gale* L., *M. cerifera* L., Europe, North America. (Material examined: *Massachusetts:* Barr 3098, MASS).

11. **Cryptodiaporthe aubertii** var. **comptoniae** (Schw.) Wehm., The Genus Diaporthe, p. 203. 1933.

Ascospores 9.5-13 × 3.5-5 μm; in branches of *Comptonia peregrina* (L.) Coult. and *Vaccinium angustifolium* Ait., eastern North America. Conidial state: *Neobarclayella flagellifera* (Ell. & Ev.) Kunze. (Material examined: *Massachusetts: Comptonia*, Barr 2325, MASS; *Michigan: Vaccinium*, Barr 999, MASS).

12. **Cryptodiaporthe phomaspora** (Cooke & Ellis) Barr, comb. nov.

Valsa phomaspora Cooke & Ellis, Grevillea **6**: 10. 1877.

Ascospores 8-11 × 3-4 μm; in branches of *Myrica cerifera* L., eastern North America. (Material examined: *Maine:* Barr 4788a, MASS; *New Jersey:* Ellis, N.A.F. 179, MASS; *Georgia:* GA 6311, GA).

13. **Cryptodiaporthe vepris** (DeLacr.) Petrak, Ann. Mycol. **32**: 445. 1934. FIG. 97

Ascospores 6.5-10 × 1.5-2.5 μm; in branches of *Rubus* spp., *Sambucus pubens* Michx., Europe, eastern North America. (Material examined: *Quebec:* Barr 1855, MASS; *Massachusetts:* Shear 3938, BPI; *New Jersey:* Ellis, N.A.F. 877, MASS; *Tennessee:* GA 6313, GA; *Georgia:* GA 6312, GA, all on *Rubus* spp.; on *Sambucus*, *Massachusetts:* Barr 2976, MASS).

The other species, which Wehmeyer (**1933**) included in *Cryptodiaporthe* and which have not been assigned previously to a section, may be disposed of in different fashions. Of the European species not known from North America: *C. lebiseyi* (Desm.) Wehm. appears to belong in section *Cryptodiaporthe*, related to *C. petiolophila* but differing in very narrow beaks; *C. niesslii* (Kunze) Wehm., also on *Acer*, should be a species of section *Cryptodiaporthe; C. robergeana* (Desm.) Wehm. belongs in section *Latispora*. *Cryptodiaporthe hranicensis* (Petrak)

Wehm. is the type species of *Amphiporthe*, and *C. castanea* (Tul.) Wehm. also appears to be a species of *Amphiporthe*, while *C. pyrrhocystis* (Berk. & Br.) Wehm. is better placed in *Allantoporthe*. The North American *C. densissima* (Ell.) Wehm. is synonymous with *Amphiporthe raveneliana*. *Cryptodiaporthe aculeans* (Schw.) Wehm. is removed to *Amphiporthe*. *Cryptodiaporthe paulula* (Cooke & Ellis) Wehm. seems better returned to *Diaporthe*. The position of *C. calosphaerioides* (Ell. & Ev.) Wehm. is doubtful, as the ascospores are described as narrow cylindroid to allantoid. *Cryptodiaporthe magnispora* (Ell. & Ev.) Wehm. is surely identical with *Melanconis everhartii* Ellis (see *Macrodiaporthe*).

30. **Amphiporthe** Petrak, Sydowia **24**: 257. (1970) 1971.

Stromatic tissues forming grayish brown entostroma, bright brown or grayish truncate erumpent conic disc, compact, prosenchymatous; perithecia grouped near base of entostroma, ±oblique, beaks eccentric, erumpent through disc, converging or separate. Asci oblong ellipsoid. Ascospores hyaline, one septate, not constricted, narrowly ellipsoid or fusoid, tapering more to basal than to upper end, often with short setose appendages at both ends.

Conidial state: as locules in entostroma; conidia hyaline or light brown, one celled, ellipsoid or allantoid (two types). (Described as *Amphicytostroma* Petrak).

In branches of deciduous trees and shrubs.

Type species: *Amphiporthe hranicensis* (Petrak) Petrak.

Petrak (**1971b**) created *Amphiporthe* to accommodate *Diaporthe hranicensis*, a species which differs from the concepts of *Diaporthe* and *Hercospora* in lacking a blackened marginal zone, from *Cryptodiaporthe* in possessing a well-developed entostroma, and from *Melanconis* in more distinctly conic stromatic disc, more numerous asci which float free in the centrum at maturity, and narrow ascospores, according to his key, which also notes the differing conidial states of these genera. Wehmeyer (**1933**) noted that *Cryptodiaporthe hranicensis*, his disposition of the species, was similar in many respects to *C. aculeans* and to *Diaporthe leiphaemia* and its variety *raveneliana*; he observed that the latter species were atypical in their respective genera. These species all seem better accommodated in *Amphiporthe* than in *Diaporthe* or *Cryptodiaporthe*.

1. **Amphiporthe hranicensis** (Petrak) Petrak, Sydowia **24**: 257. (1970) 1971.

Petrak (**1914**) provided a full description of this species, on *Tilia* spp. in Europe, and Wehmeyer (**1933**) described and illustrated it when he transferred it to *Cryptodiaporthe*. The stromatic disc is pulverulent grayish brown, and the nonappendaged ascospores measure 11-15 × 2.5-4 µm. The conidial state was described as *Amphicytostroma tiliae* (Sacc.) Petrak (**1921**). *Amphicytostroma* Petrak (**1919**) was based upon the formation of conidial locules in a ring around central stromatic tissue, into which perithecial beaks soon protrude.

2. **Amphiporthe raveneliana** (Thümen & Rehm) Barr, comb. nov.

Diaporthe raveneliana Thümen & Rehm, Flora **61**: 178. 1878.

Diaporthe leiphaemia (Fries) Sacc. var. *raveneliana* (Thümen & Rehm) Wehm., The Genus *Diaporthe*, p. 178. 1933.

Cryptodiaporthe raveneliana (Thümen & Rehm) Kobayashi, Bull. Gov. Forest Exp. Sta. **226**: 54. 1970.

Diaporthe densissima Ellis, Amer. Nat. **17**: 316. 1883.

Cryptodiaporthe densissima (Ellis) Wehm., The Genus *Diaporthe*, p. 198. 1933.

This fungus develops in branches of *Quercus* spp. in North America and Japan. The disc is brownish, ascospores are 12-15(-18.5) × 3-5 (-6) µm, appendaged at both ends, and the conidial state is *Fusicoccum*-like. Two types of conidia are formed, ellipsoid and ±allantoid, and according to Kobayashi (**1970**), both are germinable. *Diaporthe densissima*, according to authentic specimens (Ellis, N.A.F. 1192, MASS) is identical with *A. raveneliana*, as Kobayashi (**1970**) believed. The European *Diaporthe leiphaemia* (Fries) Sacc. probably also belongs in *Amphiporthe*. It differs from *A. raveneliana* in more erumpent stromatic discs and in longer, more strongly curved and constricted ascospores.

3. **Amphiporthe aculeans** (Schw.) Barr, comb. nov. FIG. 98

Sphaeria aculeans Schw., Syn. Amer. Bor. n. 1399. 1832.

Wehmeyer (**1933**) provided synonymy of this species. This fungus develops in species of *Rhus* in North America and Japan and has a bright brownish disc., whitish pulverulent at times, narrow ascospores (11-)15-8(-20) × 2.5-3.5 μm. The setose appendages are 2.5-4 μm long. The conidial state, according to Wehmeyer (**1933**) is *Phomopsis*-like, with ovoid or oblong-fusoid conidia. In addition, Wehmeyer observed sporodochial structures in culture; these produced ovoid or fusoid, blackish, one-celled conidia. A Massachusetts collection (Barr 6271, MASS) on *Rhus glabra* L., bears the conidial state in locules which are ±ring-like around the central stromatic mass. The oblong or ellipsoid, one-celled conidia are light brown in mass and measure 5-9 × 2.5-3.5 μm. Perithecial beaks and immature perithecia are present with the conidial state. Mature perithecia and beaks obliterate the conidial locules, although some conidia are still visible. Older stromata have elongate "sporodochia" producing brown, one-celled conidia, 7-8 × 3-3.5 μm, from the inflated tips. The hyphal structure of sporodochia is similar to that of the perithecial beaks. Similar structures are found at times emerging from the periderm.

Cryptodiaporthe castanea (Tul.) Wehm., on *Castanea* spp. in Europe and Japan (Kobayashi **1970**) also appears to belong in *Amphiporthe* with its well developed ectostromatic disc, entostroma, and similar conidial state. **Amphiporthe castanea** (Tul.) Barr, comb. nov. has as basionym *Sphaeria castanea* Tul., Ann. Sci. Nat. Bot. Ser. III, **15**: 379. 1851. The disc is brown and the short appendaged ascospores 11-16 × 2-3.5(-5.5) μm.

Diaporthe utahensis (Sacc.) Petrak, Ann. Mycol. **22**: 80. 1924 (*Apioporthe* in the sense of Wehmeyer, **1933**), from description and illustration could well belong in *Amphiporthe*. The stromatic disc is light colored, ascospores 11-13 × 2.5-4 μm, bent, with the lower cell short than the upper.

31. **Cryphonectria** (Sacc.) Sacc., Syll. Fung. **17**: 783. 1905.

Nectria subgenus *Cryphonectria* Sacc., Syll. Fung. **2**: 507. 1883.

Stromatic tissues prosenchymatous, entostroma mixed with substrate cells, ectostromatic disc erumpent, compact, usually brightly colored, yellow, orange or bright brownish; perithecia grouped, ±oblique, beaks elongate, converging through stromatic disc. Asci ellipsoid. Ascospores hyaline, ellipsoid or ovoid, one septate.

Conidial state: *Endothiella*-like or *Dendrophoma*-like: locules one or several in stroma; conidiophores simple or branched, conidiogenous cells phialidic; conidia hyaline, oblong, cylindroid or allantoid, one celled, minute.

Parasitic or saprobic in branches of deciduous trees.

Type species: *Cryphonectria gyrosa* (Berk. & Br.) Sacc.

Cryphonectria is separated from *Endothia* in the present scheme, and the two genera are placed in different families on the bases of stromatic configuration and ascospore characters: *Cryphonectria* has a valsoid configuration of perithecia in prosenchymatous stroma and ellipsoid or ovoid one-septate ascospores, while *Endothia* has a diatrypoid configuration of perithecia in pseudoparenchymatous stroma and allantoid, one-celled ascospores. When Saccardo (**1883**) described *Cryphonectria* as a subgenus of *Nectria*, he included several species. Von Höhnel (**1909a**) designated *C. gyrosa* as type of the genus and synonymized it with *Endothia*. The name *Endothia gyrosa* (Berk. & Br.) v. Höhnel is illegitimate because of the older *Endothia gyrosa* (Schw.) Fries, but *Cryphonectria gyrosa* is a valid name.

1. **Cryphonectria gyrosa** (Berk. & Br.) Sacc., Syll. Fung. **17**: 784. 1905.

Diatrype gyrosa Berk. & Br., J. Linn. Soc. London **14**: 124. 1875.

Nectria gyrosa (Berk. & Br.) Berk. & Br., J. Linn. Soc. London **15**: 86. 1877.

Endothia tropicalis Shear & Stevens, USDA Bull. **380**: 20. 1917.

Perithecia 300-440 μm diam., in brightly colored stroma. Asci 21-26 × 5-6 μm. Ascospores 3.5-7 × 2-2.5 μm. Conidial state: *Dendrophoma*-like; conidia 2.5-3.5 × 1.5-2 μm.

In branches of deciduous trees in tropical regions; described originally from Ceylon on *Elaeagnus* sp.

Material examined: *Brazil:* on *Eucalyptus grandis* Sm., Dionisio, Minas Gerais, 2 Feb 1973, C.S. Hodges (MASS).

Kobayashi (**1970**) described and illustrated *Endothia havanensis* Bruner and included *C. gyrosa* as a synonym. However, *E. havanensis* according to the type description and to Kobayashi's measurements has larger ascospores as well as conidia. The ascospores measure 7.5-12 × 3-4 μm and the conidia 3-4.5 × 1-1.5 μm. This species is known on a number of deciduous trees from Cuba and Japan. It agrees with the concept of *Cryphonectria*, therefore, the combination of **C. havanensis** (Bruner) Barr, comb. nov. is proposed. (Basionym: *Endothia havanensis* Bruner, Mycologia **8**: 241. 1916.)

The chestnut blight fungus, **Cryphonectria parasitica** (Murr.) Barr, comb. nov. (FIG. 99) (Basionym: *Diaporthe parasitica* Murrill, Torreya

6: 189. 1906), is similar in morphology to *C. gyrosa*. The virulently parasitic nature of *C. parasitica* on American chestnut is one of the distinguishing characters of the species. In addition ascospores in this species are 7-11 × 3.5-5.5 μm, more broadly ellipsoid than in the two preceding species. Anderson (**1914**) and Shear et al. (**1917**), among others, detailed studies on development and pathogenicity of the fungus.

Cryphonectria radicalis (Schw. ex Fries) Barr, comb. nov. (Basionym: *Sphaeria radicalis* Schw. ex Fries, Elenchus Fung. **2**: 73. 1828), on a number of hardwood trees, including *Castanea*, in North America, Europe, and Japan, is saprobic. The ascospores are small, 5.5-10 × 2.5-4.5 μm.

Endothia nitschkei Otth is the older name for *Endothia japonica* Kobayashi & Ito, as Kobayashi (**1970**) pointed out. The asci and ascospores are larger than in the previous species, with the ascospores measuring 8.5-17 × 3-5.5 μm. The species is transferred to **Cryphonectria nitschkei** (Otth) Barr, comb nov. (Basionym: *Endothia nitschkei* Otth, Mitt. Nat. Ges. Bern **8**. 1868). **Cryphonectria macrospora** (Kobayashi & Ito) Barr, comb. nov. (Basionym: *Endothia macrospora* Kobayashi & Ito, Ann. Phytopath. Soc. Japan **21**: 152. 1956), has still larger ascospores, 14-18 × 5-5.5 μm. *Endothia tetraspora* Kobayashi (Trans. Mycol. Soc. Japan **6**: 99. 1965), has large ellipsoid-fusoid, thick-walled, one-celled ascospores, 21-31 × 7.5-10 μm, and belongs in *Cryptosporella*.

32. **Ophiovalsa** Petrak, Sydowia **19**: 272. (1965) 1966.

Cryptospora Tulasne & Tulasne, Sel. Fung. Carp. **2**: 144. 1863, non Karelin & Kirilow, 1842.

Stromatic tissues prosenchymatous, forming a small extostromatic disc above perithecia, entostroma scanty, as hyphae around perithecial walls; perithecia circinate, beaks converging and erumpent together through disc. Asci ellipsoid. Ascospores hyaline, elongate, allantoid or cylindroid or ends enlarged, one celled or one to several septate.

Conidial state: *Disculina* v. Höhnel: locules acervulus-like in ectostroma; conidia hyaline, often falcate, curved, one celled.

In branches of deciduous trees and shrubs.

Type species: *Ophiovalsa suffusa* (Fries) Petrak.

The species of *Ophiovalsa* could be derived from those of *Valsa* section *Circinatae*, where entostromatic tissues are reduced and extostromatic disc is well developed, by elongation and often by the formation of septa in the ascospores.

Petrak (**1966a**) erected subgenus *Ophiovalsella* to accommodate *Ophiodothella panamaensis* Stevens. This fungus deviates from *Ophiovalsa* in centrum and in ascus structure and is retained in *Ophiodothella.*

1. **Ophiovalsa suffusa** (Fries) Petrak, Sydowia **19**: 272. (1965) 1966.

Sphaeria suffusa Fries, Syst. Mycol. **2**: 399. 1823.

Valsa suffusa (Fries) Fries, Summa Veget. Scand. p. 412. 1849.

Cryptospora suffusa (Fries) Tulasne & Tulasne, Sel. Fung. Carp. **2**: 145. 1863.

Stromata forming small yellowish brown disc; perithecia circinate, 350-500 μm diam, beaks elongate, converging and usually obliterating stromatic disc. Asci 45-90 × 16-25 μm. Ascospores 30-50 × 3-5 μm, one celled.

Conidial state: *Disculina neesii* (Corda) v. Höhnel: conidia 21-30 × 3.5-5 μm, one celled, curved.

In branches of *Alnus* spp., cosmopolitan.

Reported from Iowa (Gilman, Tiffany, and Lewis, **1959**).

Ophiovalsa femoralis (Peck) Petrak (Sydowia **19**: 273. (1965) 1966) (FIG. 100) is more common in eastern North America than *O. suffusa.* The ascospores of this species are thickened toward each end to 5.5-6.5 μm wide, while the mid portion is 3.5-4.5 μm wide. A median septum is usually formed.

Several species of *Ophiovalsa* in addition to *O. suffusa* have one-celled ascospores: *O. betulae* (Tul.) Petrak, *O. corylina* (Tul.) Petrak, *O. tiliae* (Tul.) Petrak, and *O. tomentella* (Peck) Petrak. Other species such as *O. cinctula* (Cooke & Peck) Petrak and **O. trichispora** (Cooke & Peck) Barr, comb. nov. (Basionym: *Valsa trichispora* Cooke & Peck in Peck, New York State Museum Rep. **29**: 50 (for 1875) 1878; type-Greenbush, New York, Nov, C.H. Peck, NYS) have ascospores which become one to three septate.

II-2c. **Valsaceae** tribe **Diaportheae** Barr, tribus nov.

Perithecia aggregata, raro singulara, stromata circumscriptiones per zonas nigrifactos marginales; in ramorum arboribus et frutibus vel herbarum caulis. Genus typicus: *Diaporthe* Nits.

Key to Genera of Tribe Diaportheae

1. Ascospores one celled, allantoid, or straight and cylindroid; stromata isolated, delimited by black conceptacle 2
1. Ascospores one septate, ellipsoid, fusoid, or cylindroid; stromata effused, or pustulate effused, or isolated, delimited by blackened marginal zone at substrate surface, ventral blackened marginal zone present or lacking .... 35. *Diaporthe*
   2. Asci octosporous or less .... 33. *Leucostoma*
   2. Asci polysporous .... 34. *Valsella*

33. **Leucostoma** (Nits.) v. Höhnel, Ber. Deutsche Bot. Ges. **35**: 637. 1917; Ann. Mycol. **16**: 134-135. 1918.

Valsa subgenus *Leucostoma* Nits.

Stromatic tissues prosenchymatous or pseudoparenchymatous, forming white or grayish brown ectostromatic disc, delimited by darkened conceptacle of compact tissues; perithecia oblique, few to numerous, beaks converging and erumpent through disc. Asci ellipsoid. Ascospores hyaline, cylindroid to allantoid, one-celled.

Conidial state *Leucocytospora* (v. Höhnel) v. Höhnel (=*Cytospora*, Sutton, 1977): conidial locules formed in stromata similar to perithecial stromata; conidiogenous cells phialidic; conidia hyaline, cylindroid to allantoid, one celled.

In branches of deciduous trees, shrubs and gymnosperms.

Type species: *Leucostoma massariana* (de Not.) v. Höhnel.

Leucostoma is regarded by a number of investigators as a subgenus of *Valsa*, but by others is held to be generically different. The compact darkened conceptacle delimiting each stroma is quite distinctive. *Leucostoma* seems to be more closely related to *Diaporthe* than to *Valsa*. The genus is illustrated by *Leucostoma kunzei* (Fries) Defago (FIG. 101).

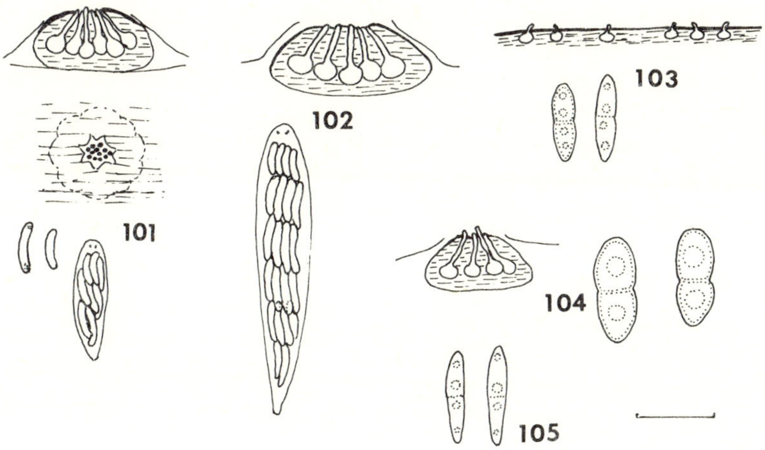

FIGS. 101-105. FIG. 101. *Leucostoma kunzei,* stroma with perithecia in section, habit, ascus, ascospores. FIG. 102. *Valsella fertilis,* stroma with perithecia in section, ascus. FIGS. 103-105. Species of *Diaporthe.* 103. *D. eres,* stroma with perithecia, ascospores. 104. *D. tuberculosa,* stroma with perithecia, ascospores. 105. *D. prunicola,* ascospores. Scale =1 mm or 10 μm.

Kern (**1955, 1957, 1961**) recognized two groups of species in *Leucostoma:* in one the stromatic disc is whitish or white, and in the other it is grayish brown to dark brown. The thickness and shape of the conceptacle, host plant, and ascospore size range all serve to delimit the groups of species. Physiological studies have been conducted on a number of species by Kern and by Hubbes (**1960a**) and others.

34. **Valsella** Fuckel, Symb. Mycol. p. 203. 1870.

Similar to *Leucostoma* in prosenchymatous stroma, ectostromatic disc, darkened conceptacle, immersed grouped perithecia. Differing in polysporous asci. Ascospores hyaline, allantoid, one celled.

Conidial state: *Cytospora,* similar to *Leucostoma* conidial states.

In branches of deciduous trees and shrubs and in gymnosperms.

Type species: *Valsella salicis* Fuckel =*V. fertilis* (Nits.) Sacc.

Petrak (**1919, 1921, 1970**, and other notes) and Shear (**1943**), did not believe that the polysporous condition was sufficient to separate the species of *Valsella* from those of *Leucostoma.* Certainly many similarities in configuration of stroma, host plants, and ascospore size ranges may be found, but proof of the existence of octosporous or polysporous strains within a species has not been demonstrated. In fact, Kern (**1961**) and Hubbes (**1960a**) have observed that within an isolate the polysporous condition is constant. Further studies are required to substantiate the generic position of *Valsella.*

Valsella fertilis (Nits.) Sacc., Syll. Fung. **1**: 161. 1882 (FIG. 102), has a whitish ectostromatic disc, blackened conceptacle, numerous small perithecia, 250-300 μm diam. The asci are polysporous with 20 or more per ascus, and ascospores measure 5.5-8 × 1-1.5 μm. It is common in branches of *Salix* spp., Europe, North America (Bartholomew, Fungi Col. 3200, MASS, is an example). *Valsella nigroannulata* Fuckel, Symb. Mycol. Nachtr. **1**: 29. 1872, represents the larger-spored series of species in *Valsella.* This species also occurs on *Salix* branches and has a whitish ectostromatic disc. The asci contain about 16 ascospores which measure 8-13 × 1.5-2.5 μm (Maine: Barr 3610a, MASS).

35. **Diaporthe** Nitschke, Pyren. Germ. p. 240. 1870.

Chorostate (Sacc.) Trav., Flora Ital. Crypt. Pars **I**: 190. 1906.

Diaporthe subgenus *Chorostate* Sacc., Syll. Fung. **1**: 607. 1882.

Septomazzantia Theissen & Sydow, Ann. Mycol. **13**: 193. 1915.

Clypeorhynchus Kirschst., Ann. Mycol. **34**: 190. 1936.

Melanoporthe Wehm., Rev. Mus. de La Plata **2**: 84. 1938.

Stromatic tissues composed of prosenchymatous hyphae, mixed with substrate tissues in entostromatic region, often forming dark marginal zones above, in epidermal tissues, and frequently deep in wood beneath perithecia or groups of perithecia; perithecia grouped or separate, oblique or lateral, beaks lateral or oblique, converging when perithecia grouped, curved when perithecia separate. Asci ellipsoid. Ascospores hyaline, rarely faintly colored, ellipsoid, fusoid or cylindroid, one septate, septum usually median, appendages short and narrow when formed.

Conidial state: *Phomopsis*: pycnidial stromata containing one or more locules, conidiogenous cells phialidic; conidia hyaline, two types formed — *a* fusoid and one celled, *β* elongate, usually curved, one celled or septate.

In woody branches of trees and shrubs, or herbaceous stalks.

Type species: *Diaporthe eres* Nits.

Chorostate (Sacc.) Trav. was erected for those species with valsoid aspect of stroma and circinate perithecia. Traverso did not specify a type species, but illustrated *C. italica* (Sacc.) Trav. Wehmeyer (**1933**) was unable to study this species and suggested that it was probably *D. medusaea* Nits. on *Populus*. Traverso emended *Diaporthe* to include those species with an effused stroma, blackened marginal zones, and perithecia erumpent separately.

Septomazzantia was based on *S. epitypha* (Cooke) Theissen & Sydow, which according to Petrak (**1954**) is a typical species of *Diaporthe*. *Clypeorhynchus* was shown to be identical with *Diaporthe* by von Arx and Müller (**1954**). *Melanoporthe*, with the type species *M. talae* (Speg.) Wehm., was separated upon the brown pigmentation in the mature, broadly elliptic ascospores. The color variation of mature spores is not of generic significance in this family, and in other respects *Melanoporthe* conforms with *Diaporthe*.

Müller and von Arx (**1962**) included several other genera in the synonymy of *Diaporthe*; these are not considered in the present study to be synonyms. *Allantoporthe* Petrak is recognized as a distinct genus, with upright perithecia and central beaks. *Clypeoporthe* v. Höhnel is also recognized as a separate genus. *Stigmatopsis* Trav. (**1906**), according to the description and illustrations, may be a genus of the Diatrypaceae. *Skottsbergiella* Petrak (in Keissler, Nat. Hist. Juan Fernandez and Easter Island **2**: 481. 1927), appears to be a distinct entity, with large perithecia immersed in a massive pseudoparenchymatous stroma, crustose externally. Petrak (**1971d**) upheld the genus which he compared in stromatic consistency to that of eutypoid fungi. *Skottsbergiella* perhaps belongs in the family Gnomoniaceae, related to *Diaporthella*.

Wehmeyer (**1933**) recognized two sections of *Diaporthe* in his monographic treatment: *Effusae*, in which dorsal blackened zones do not penetrate into bark tissues, and *Pustulatae* in which the dorsal blackened zones dip into the bark between perithecial groups. Munk (**1957**) also grouped the species of *Diaporthe* which he considered on the basis of types of stromatic configuration. The type of stroma and the presence of blackened zones is not always obvious in individual collections. For primary separation of groups of species, I prefer to utilize major difference in ascospore morphology. For the North American species of *Diaporthe*, the following arrangement of sections and groups of species is proposed.

Diaporthe section **Diaporthe**

Ascospores narrow, ±fusoid ellipsoid, often inequilateral or slightly curved, with two globules per cell. Type species: *Diaporthe eres* Nits. FIG. 103.

Group of *D. eres:* usually in woody branches; perithecia erumpent separately, entostroma effused, not limited.

Group of *D. arctii* (Lasch) Nits.: usually in herbaceous stalks; perithecia erumpent separately, entostroma effused, not limited.

Group of *D. pardalota* (Mont.) Fuckel: in herbaceous stalks or soft woody branches; perithecia erumpent separately, entostroma sharply margined, isolated.

Group of *D. strumella* (Fries) Fuckel: in woody branches; perithecia usually grouped and erumpent together, entostroma pustulate effused.

Diaporthe section **Latispora** Barr, sectio nov.

Ascosporae latae, ellipsoideae vel fusoidae, rectae, plerumque globulo uno intra cellula. Species typica: *D. staphylina* Ell. & Ev.

Group of *D. staphylina*: usually in woody branches; perithecia erumpent separately, entostroma effused or somewhat limited.

Group of *D. tuberculosa* (Ell.) Sacc. (FIG. 104): usually in woody branches; perithecia grouped and erumpent together, entostroma pustulate effused.

Diaporthe section **Cylindrica** Barr, sectio nov.

Ascospore angustae, cylindroidae vel fusoidae elongatae, rectae vel curvae, globulis duobus vel ultra. Species typica: *D. prunicola* (Peck) Wehm. FIG. 105.

Of the species monographed by Wehmeyer (**1933**), *D. decedens* (Fries) Fuckel is considered to belong in *Allantoporthe*, as does *D. tessella* (Pers. ex Fries) Rehm. *Diaporthe leiphaemia* var. *raveneliana* is transferred to *Amphiporthe*. *Diaporthe tiliacea* (Ellis) v. Höhnel has proved troublesome to place, but its affinities appear to lie with *Phragmodiaporthe*.

B. **Melanconidineae** Barr, subordo nov.

Perithecia amplitudine media vel magna, peridiis tenuibus vel crassis, rostris latis; asci cylindroidi, oblongi, vel ellipsoidi, fundis plerumque persistentibus affixis in subhymenio; ascosporae hyalinae, flavobrunneae, vel fuscae, pariete firmo plerumque incrassato. Genus typicus: *Melanconis* Tul. (2 families).

III. Family **Pseudovalsaceae** Barr, fam. nov.

Perithecia erecta, rostris centralis, raro eccentricis, erumpentis seorsim. Genus typicus: *Pseudovalsa* Ces. & de Not. (2 subfamilies).

Key to Subfamilies and Tribes of Pseudovalsaceae

1. Perithecia grouped, stromatic tissues prosenchymatous, forming compact disc at surface of substrate........................ Subfamily Allantoporthoideae, p. 151
   One tribe, Allantoportheae
1. Perithecia single or grouped, stromatic tissues pseudoparenchymatous, at times reduced to clypeus or disc at surface of substrate....... Subfamily Pseudovalsoideae . 2
   2. Perithecia single, stromatic tissues adhering firmly to peridium or forming flared clypeus at surface of substrate....................... Tribe Ditopelleae, p. 155
   2. Perithecia grouped, stromatic tissues forming compact disc at surface of substrate, more diffuse laterally, often delimited by blackened ventral zones................ .... Tribe Pseudovalseae, p. 163

III-1. Subfamily **Allantoporthoideae** Barr, subfam. nov.

Perithecia aggregata, stromata contexta condensata, instar disci stromatici; in ramorum arboribus et frutibus. Genus typicus: *Allantoporthe* Petrak.

Key to Genera of Subfamily Allantoporthoideae

1. Entostroma slightly developed in periderm, containing few large perithecia; ascospores one septate; conidia allantoid, short and narrow, one celled .... 36. *Allantoporthe*
1. Entostroma well developed in periderm, containing numerous medium large perithecia; ascospores one to several septate; conidia elongate, fusoid clavoid to cylindroid, usually several septate .... 38. *Phragmodiaporthe*

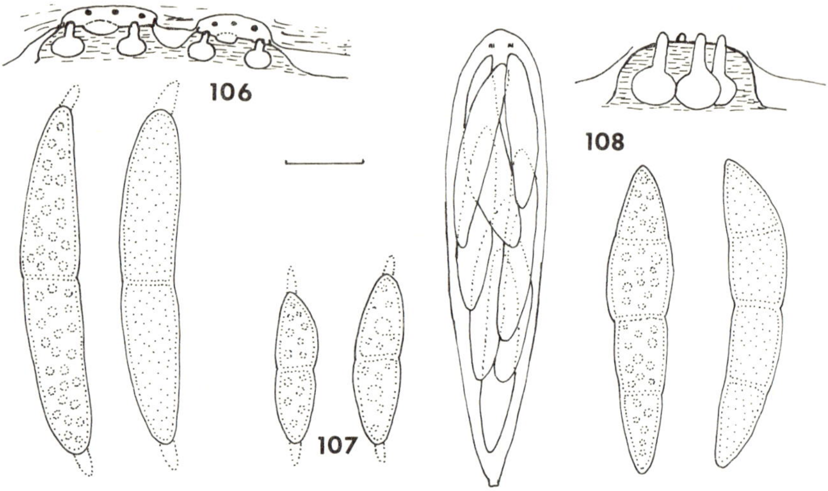

FIGS. 106-108. FIGS. 106, 107. Species of *Allantoporthe.* 106. *A. tessella,* stroma with perithecia in section, ascospores. 107. *A. decedens,* ascospores. FIG. 108. *Phragmodiaporthe caryae,* stroma with perithecia in section, ascus, ascospores. Scale =1 mm or 10 μm.

III-1a. Tribe **Allantoportheae**

36. **Allantoporthe** Petrak, Hedwigia **62**: 288. 1921.

Stromatic tissues prosenchymatous, forming a blackened disc and blackened marginal zones; perithecia ±circinate, upright, beaks separately erumpent. Asci broadly oblong or ellipsoid. Ascospores cylindroid fusoid or ellipsoid, elongate, hyaline, often curved, tapered to ends, one septate, contents homogeneous, minutely guttulate, wall smooth, appendage short and narrow.

Conidial state: formed in cavities in stromatic locules; conidia allantoid, hyaline, one celled.

In branches of deciduous trees and shrubs.

Type species: *Allantoporthe tessella* (Pers. ex Fries) Petrak.

Wehmeyer (**1933**) reduced Petrak's genus to synonymy with *Diaporthe.* In **1971a** Petrak reaffirmed the validity of *Allantoporthe*; in its diagnostic characters this genus is quite distinct from *Diaporthe.*

Cytomelanconis Naumov (Notulae Syst. Sect. Crypt. Inst. Bot. Acad. Sci. USSR **7**: 108. 1951), according to the brief description in Müller and von Arx (**1962**), as Anhang to Diaporthaceae, may well belong here. The conidial state was described as belonging to *Cytospora*; ascospores in the type and only species, *C. systema-solare* Naumov, on *Salix* sp., are fusoid, slightly curved, one septate, 50-64 × 8-10 μm, two per ascus.

1. **Allantoporthe tessella** (Pers. ex Fries) Petrak, Hedwigia **62**: 289. 1921. FIG. 106

See Wehmeyer (**1933**) for synonymy, under *Diaporthe tessella.*

Stromata 2-3 mm diam, depressed, mid portion blackened as ectostromatic disc; perithecia 660-770 μm diam, 385-550 μm high, beaks broad, erumpent to surface of stroma. Asci 90-140 × 20-25 μm. Ascospores 40-60 × 7-9(-11) μm, appendages ca. 3-4 μm long, 2 μm wide.

In branches of *Salix* sp., Europe, North America.

Material examined: *North America: Ontario:* Ell. & Ev. N.A.F. 2523; *New York:* New York Fungi, C.L. Shear 172 (as *Melanconis salicina* Ell. & Ev.) (both MASS).

Other species with almost diatrypelloid stromatic grouping, large perithecia with central, separately erumpent beaks around an eroded blackened disc, and conidia deviating from those of *Diaporthe*, should also be transferred to *Allantoporthe.* Such a species is *Diaporthe decedens* (Fries) Fckl.

2. **Allantoporthe decedens** (Fries) Barr, comb. nov. FIG. 107

Basionym: *Sphaeria decedens* Fries, Syst. Mycol. **2**: 481. 1823.

See Wehmeyer (**1933**) for synonymy, under *Diaporthe decedens.* The stromatic groups at times are marked by a dark line at the sides, but not normally; this species differs from *A. tessella* particularly in smaller asci, 64-70 × 10-12 μm, and ascospores, 15-22 × 4-6 μm. The ascospores, as in those of *A. tessella*, have a short blunt appendage at each end and contain numerous small guttules.

In branches of *Corylus* spp., Europe, North America.

Material examined: *North America: Ontario:* Ell. & Ev. N.A.F. 2929; *Iowa:* Ell. & Ev. N.A.F. 2818. (both MASS).

From the descriptions and illustrations (Wehmeyer, **1933**) of the European *Cryptodiaporthe pyrrhocystis* (Berk. & Br.) Wehm., this species appears to belong in *Allantoporthe.*

31. **Phragmodiaporthe** Wehmeyer, Mycologia **33**: 54. 1941.

Stromatic tissues prosenchymatous, compact, blackened above as ectostromatic disc, dipping down as blackened lateral zone at sides of entostroma, brownish within, substrate cells included; perithecia grouped, upright, beaks central, erumpent closely together but not truly converging, relatively short and broad, not extending far beyond stroma surface. Asci clavoid oblong, apical annulus indistinct. Ascospores hyaline to yellowish, becoming brown in age at times, elongate fusoid, tapering to rounded or pointed ends, several septate.

Conidial state: in cavities in stroma (*Cytosporina*-like), or on surface of stroma (*Coryneum* Nees ex Schw.), or beneath periderm in minute cavities; conidia hyaline or brown, elongate, fusoid clavoid to cylindroid, nonseptate to several septate.

In branches of deciduous trees.

Type species: *Phragmodiaporthe caryae* (Peck) Wehmeyer.

Phragmodiaporthe is evidently related to *Allantoporthe;* the several-septate ascospores and the different conidial states distinguish the two genera. The stromatic tissues are rather light brown and marked above and at the sides by blackened zones in *P. caryae*, much as in *A. tessella*. *Phragmodiaporthe aculeata* (Petch) Wehm. was not based upon Petch's type specimen, but on a later collection. The species does not appear to be congeneric with *P. caryae*, rather it seems close to *Ophiovalsa*.

1. **Phragmodiaporthe caryae** (Peck) Wehm., Mycologia **33**: 55. 1941.
FIG. 108

Cryptospora caryae Peck, New York State Museum Rep. **38**: 106. 1885.

Winterella caryae (Peck) Kuntze, Rev. Gen. Pl. Pars **II**: 34. 1891.

Pseudovalsa fairmani Ell. & Ev., Proc. Rochester Acad. for **1890**: 51.

Aglaospora fairmani (Ell. & Ev.) Kuntze, Rev. Gen. Pl. **3**(3): 441. 1898.

Valsa apatela Ell. & Holway, Bull. Lab. Nat. Hist. Univ. Iowa **3**: 42. 1895.

Calospora apatela (Ell. & Holway) Sacc. & Sydow in Sacc., Syll. Fung. **14**: 593. 1899.

Stromata blackened at surface and sides, brown internally; perithecia 300-500 μm diam, 300-450 μm high, beak up to 400 μm long, 250-280 μm wide. Asci 75-110 × 13-16.5 μm. Ascospores 29-44 × (5-)6.5-8 μm, hyaline, straight or slightly curved, one to three septate.

Conidia in irregular cavities in stroma; conidia 35-43 × 3.5-5.5 μm, hyaline, nonseptate or up to three septate.

In branches of *Carya alba* (L.) K. Koch.

Material examined: *New York:* Knowersville, May 1884, C.H. Peck, type, NYS).

Diaporthe tiliacea (Ellis) v. Höhnel, which is misplaced in *Diaporthe*, seems better disposed of in *Phragmodiaporthe*, although its conidial state is produced in minute cavities beneath the host periderm (Wehmeyer, **1923**, as *Diaporthe furfuracea*).

2. **Phragmodiaporthe tiliacea** (Ellis) Barr, comb. nov.

Basionym: *Diatrype tiliacea* Ellis, Amer. Nat. **17**: 195. 1893.

Wehmeyer (**1933**) provided synonymy, description, and illustrations of this species as *Diaporthe tiliacea.*

III-2. **Pseudovalsaceae** subfamily **Pseudovalsoideae**

Perithecia single or grouped in well-defined pseudoparenchymatous stromatic tissues, or stromatic tissues at times reduced to pseudoparenchymatous clypeus or disc (3 tribes).

III-2a. **Ditopelleae** Barr, tribus nov.

Perithecia singulara, stromata contexta adhaerentes perithecia vel clypeata; in ramorum vel foliis arboribus et frutibus, herbarum caulis, vel monocotyledonibus culmis. Genus typicus: *Ditopella* Schroeter.

Key to Genera of Tribe Ditopelleae

1. Stromatic tissues closely surrounding individual perithecia 2
1. Stromatic tissues flaring out as clypeus around base of beak 3

2. Ascospores one celled .... 38. *Dicarpella*
2. Ascospores one septate .... 39. *Sydowiella*
3. Ascospores one septate, asci polysporous .... 40. *Ditopella*
3. Ascospores three septate, asci octosporous .... 41. *Phragmoporthe*

Anisomyces Theissen & Sydow (Ann. Mycol. **12**: 270. 1914), with the type species *A. nectrioides* (Rehm) Petrak, from tropical America, belongs in this tribe. Stromatic tissues surround the short beak; the ascospores are apiosporous and brownish at maturity.

Gibbelia Sacc. (Atti R. Ist. Venet. Sci. Venezia ser. 6, **3**: 723. 1885) has the sole species *G. dothideoides* Sacc., in wood of unknown tree from Australia. The stromatic tissues around separately immersed perithecia are pseudoparenchymatous, according to von Arx and Müller (1954). Ascospores are broadly ellipsoid, hyaline and one celled.

38. **Dicarpella** Sydow & Sydow, Ann. Mycol. **18**: 181. 1920.

Disperma Theissen, Verh. Zool.-Bot. Ges. Wien **66**: 390. 1916, non Clarke 1899.

Stromatic tissues pseudoparenchymatous, closely surrounding individual perithecia and thickening beaks; perithecia separate, upright, immersed in leaf tissue, beaks short, broad, central; peridium of several layers, brownish externally, apical tissues yellowish. Asci ellipsoid, oblong or nearly clavoid, octosporous or less. Ascospores hyaline or yellowish, ellipsoid or fusoid, tapered to ends, one celled, contents granular guttulate.

In leaves and petioles of deciduous trees.

Type species: *Dicarpella bina* (Harkn.) Sydow & Sydow.

1. **Dicarpella bina** (Harkn.) Sydow & Sydow, Ann. Mycol. **18**: 181. 1920.
FIG. 109

Physalospora bina Harkness, Bull. Calif. Acad. Sci. **1**: 43. 1884.

Disperma bina (Harkn.) Theissen, Verh. Zool.-Bot. Ges. Wien **66**: 390. 1916.

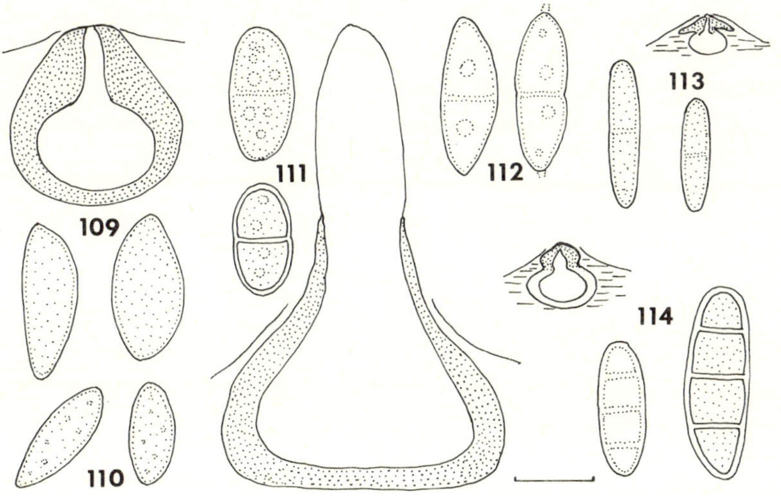

FIGS. 109-114. FIGS. 109, 110. Species of *Dicarpella*. 109. *D. bina*, perithecium in stroma in section, ascospores. 110. *D. georgiana*, ascospores. FIGS. 111, 112. Species of *Sydowiella*. 111. *S. fenestrans*, perithecium in stroma in section, ascospore. 112. *S. depressula*, ascospores. FIG. 113. *Ditopella ditopa*, habit of perithecium and stromatic clypeus, ascospores. FIG. 114. *Phragmoporthe conformis*, habit of perithecium and stromatic clypeus, ascospores. Scale =1 mm, 100 μm, or 10 μm.

Perithecia immersed in yellowish, irregular, brown-margined leaf spots, 200-220 μm diam, 230-260 μm high, beak opening epiphyllously; peridium 20-24 μm wide at base and sides, enlarged to 30-45 μm at base of beak including clypeus, yellowish. Asci 42-55 × 11-16.5 μm, two spored. Ascospores 18.5-25 × 7.5-12 μm.

In spots in leaf of *Quercus agrifolia* Née, California.

Material examined: *California:* Mt. Tamalpais, Marin Co., 20 Jan 1882 (according to packet label), 2 Jun 1882 (field label), Harkness (CAS).

2. **Dicarpella georgiana** (Miller & Thompson) Barr, comb. nov. FIG. 110

Gnomoniella georgiana Miller & Thompson, Mycologia **32**: 8. 1940.

Perithecia immersed, usually opening epiphyllously, 350-550 μm diam, 260-440 μm high, peridium 9-15 μm wide at base and sides, brown, beak and surrounding tissues up to 50 μm wide, yellowish, darkened externally, vinaceous in KOH. Asci 47-72 × 12-16 μm, eight spored. Ascospores 13-18 × 6-8 μm.

In overwintered leaves and petioles, occasionally tips of twigs, of *Nyssa* and *Liquidambar* species, southeastern North America.

Material examined: *Georgia:* 7 miles west of Harlem, 16 Mar 1939, Thompson & Miller (types, GA, BPI, NY); Towns Creek, Washington Co., 20 Mar 1939; River Road, Agric. Campus, Athens, 21 Mar 1939, Miller; 4 miles east of Pavo, 4 Jun 1943, Manning & Miller; Agric. Campus, Athens, 11 Mar 1939, Miller; 4 miles southeast of Thomson, 16 Mar 1939, Miller; on GA 6334 of *Cryptodiaporthe paulula* (all GA); Barr 6176 (MASS).

A conidial fungus is circumstantially connected with *D. georgiana.* Pycnidia are immersed in leaf tissues, opening hypophyllously, with short broad apex, globose or slighly depressed, 160-300 μm diam, 160-200 μm high; peridium ca. 20 μm wide, yellowish brown cells, vinaceous in KOH. Conidiophores lining cavity, short, 7-10 μm long. Conidia 13-22 × 6-8 μm, hyaline or yellowish, ovoid or ellipsoid, one celled, with truncate basal end, upper end bearing slender hyaline appendages, 12-15.5 × 1 μm, contents granular with one globule. This appears to be a foliicolous variant of *Harknessia americana* (Mont.) Sutton.

Similarities in conidium and ascospores shape and in reaction of the yellowish cells of pycnidium wall and beak tissues of perithecia to KOH, as well as the close association on leaves suggest the connection of these two entities.

39. **Sydowiella** Petrak, Ann. Mycol. **21**: 30. 1923.

Stromatic tissues pseudoparenchymatous, closely surrounding individual perithecia, appearing like double wall; perithecia immersed, upright, becoming partially erumpent, separate or grouped at times, beaks central, broad, short to long. Asci ellipsoid oblong or nearly cylindroid, apical annulus broad. Ascospores hyaline, broadly ellipsoid or ellipsoid fusoid, septum median, slightly constricted, contents with one or two globules, wall smooth or finely roughened, at times remnants of gelatinous appendages visible at tips.

In stalks of large herbaceous plants or in soft woody branches.

Type species: *Sydowiella fenestrans* (Duby) Petrak.

1. **Sydowiella fenestrans** (Duby) Petrak, Ann. Mycol. **21**: 30. 1923. FIG. 111

Sphaeria fenestrans Duby in Klotzsch-Rabenh. Herb. Mycol. ed. 1, n. 1933. 1855; Bot. Zeit. **13**: 285. 1885.

Gnomonia fenestrans (Duby) Sacc., Syll. Fung. **1**: 562. 1882.

Didymosphaeria fenestrans (Duby) Winter, Hedwigia **25**: 24. 1886.

Gnomonia epilobii Auersw., Mycol. Europ. **5/6**: 26. 1869.

Sphaeria epilobii Fuckel, Fungi rhen. 2020; Symb. Mycol. p. 113. 1870.

Didymella epilobii (Fckl.) Sacc., Syll. Fung. **1**: 556. 1882.

Perithecia 385-690 μm diam, 230-480 μm high, beaks 230-480 μm long, 154-205 μm wide. Asci 90-148 × 9.5-16 μm. Ascospores 15-21 × 7.5-10 μm, broadly ellipsoid, wall finely roughened at times.

In overwintered stalks of *Epilobium angustifolium* L., Europe, North America.

Material examined: *Europe:* Rehm, Ascom. 141b (MICH). *North America: Quebec:* Barr 1911, 1988 (MASS); *British Columbia:* Barr 5580a (MASS).

From the description and figures of *Gnomonia heimii* Negru (Bull. Soc. Mycol. France **80**: 223. 1964), this name is probably synonymous with *Sydowiella fenestrans*. The fungus was collected in Roumania on *Epilobium angustifolium* stalks (Negru, **1964**), has upright perithecia with stout beaks and ±cylindroid asci. The ascospores were described as slightly longer than those of *S. fenestrans*, 20-25 × 8-10 μm, but otherwise the fungus appears to be identical with *S. fenestrans*.

2. **Sydowiella depressula** (Karst.) Barr, comb. nov. FIG. 112

Gnomonia depressula Karsten, Mycol. Fenn. **2**: 121. 1873.

Perithecia 250-480(-600) μm diam, 150-400(-500) μm high, beaks 150-450(-1200) μm long, 60-180 μm wide. Asci 60-90(-135) × 11-16.5 (-22.5) μm. Ascospores 20-26 × 5.5-9(-10) μm, ellipsoid fusoid, wall smooth, remnant of gelatinous appendage at each end.

In dead branches of *Rubus* spp., Europe, North America.

Material examined: *Alaska:* Sitka, 16 Aug 1922, N.E. Stevens; 30 Jul 1923, J.P. Anderson; *British Columbia:* Ross Spur, 21 Jun 1939, G.G.

Hedgcock (all BPI); *Colorado:* Deansbury, Apr 1897, E. Bethel (Ell. & Ev. N.A.F. 3525 of *Metasphaeria rubicola* Ell. & Ev., in Dearness Herb. in DAOM, but not all specimens).

The longer, more fusoid, pointed ascospores distinguish this species from *S. fenestrans.* The exsiccati specimens of *Metasphaeria rubicola* Ell. & Ev. are apparently mixed. That in MASS agrees with the original description and is *Clathridium massarinum* (*Discostroma*). The specimen in Dearness Herbarium differs in all respects from the former and is referrable to *Sydowiella depressula.* It has asci 93-135 μm long, while in the Alaskan specimens asci are shorter. The fungus described as "*Sydowiella* sp. cf. *Didymella mesneriana*" on *Rubus* by Munk (**1957**) is probably identical with *S. depressula,* according to Munk's description and figures; only the longer asci, 100-200 μm, differ from those of the North American collections.

Munk (**1953**) transferred *Neokeissleria ribis* (P. Henn & Ploetn.) Petrak to *Sydowiella* without properly citing the basionym. I prefer to retain *Neokeissleria* as a separate genus and to exclude it from the Diaporthales. In **1957**, Munk also included *Sydowiella ambigua* (Mout.) Munk in the genus, but this species has horizontal perithecia and lateral beaks, the perithecia in small groups and the beaks converging but separate, and deviates from *Sydowiella. Melanconis ambigua* Mout. (Bull. Soc. Roy. Bot. Belg. **39**: 43. 1900) was a species which Wehmeyer (**1941b**) was unable to examine; he did not suggest what its position might be.

40. **Ditopella** Schroeter in Cohn, Kryptogamenfl. Schlesiens **3**: 388. 1897.

Stromatic tissues pseudoparenchymatous, reduced to ectostromatic disc surrounding and united with beak tissues, extending outward in periderm as a flaring clypeus; perithecia immersed separately, upright, beaks central, broad. Asci ellipsoid or elongate, polysporous. Ascospores hyaline, narrowly ellipsoid, nearly cylindroid, tapered to rounded ends, nonseptate or one (two) septate.

In branches of deciduous trees.

Type species: *Ditopella ditopa* (Fries) Schroeter.

Reid and Booth (**1967**) described and illustrated this fungus and provided the synonymy listed below. Although *D. ditopa* has not yet been reported from North America, it is included in this account because I expect that it will be collected, probably in western North America and associated with *Phragmoporthe conformis* as the two species are in Europe.

1. **Ditopella ditopa** (Fries) Schroeter, Kryptogamenfl. Schlesiens **3**: 388. 1897. FIG. 113

Sphaeria ditopa Fries, Syst. Mycol. **2**: 481. 1823.

Halonia ditopa (Fries) Fries, Summa Veget. Scand. p. 397. 1849.

Sphaeria ditopa forma *polyspora* Cooke, Handbook of British Fungi **2**: 888. 1871.

Cryptospora ditopa (Fries) Karst., Mycol. Fenn. **2**: 80. 1873.

Ditopella fusispora de Not., Sfer. Ital. p. 42. 1863.

Physalospora fusispora (de Not.) Cooke, Grevillea **17**: 88. 1889.

Cryptospora suffusa forma *minor* Tul., Sel. Fung. Carp. **2**: 145. 1863.

Ectostromatic disc ±300-350 μm thick, extending beyond perithecium, perithecia 400-600 μm diam, 300-350 μm high, beaks 200-250 μm long. Asci 45-90 × 13-22.5 μm, 16 to 24 spored. Ascospores 14.5-20 × 3-4.5 μm, nonseptate or more frequently one septate, septum median.

In branches of *Alnus* spp., Europe.

Material examined: *Europe:* Rehm: Ascom. 1139; Roumeguere Fungi sel. exs. 7145 (both NY, labelled *Metasphaeria conformis*, and bearing both *D. ditopa* and *Phragmoporthe conformis*).

Ditopella hosackiae (Cooke & Harkn.) Sacc., Syll. Fung. **1**: 451. 1882. (*Sphaerella ? hosachiae* Cooke & Harkn. Grevillea **9**: 86. 1881), according to the type collection in CAS bears only *Microdiplodia hosackiae* (Cooke & Harkn.) Bonar. Bonar (**1965**) collected a species of *Mycosphaerella* mixed in with the *Microdiplodia*. This was similar to the incomplete description of *Sphaerella ? hosackiae*, and Bonar transferred the species as *Mycosphaerella hosackieae* (Cooke & Harkn.) Bonar, Mycologia **57**: 384. 1965, and provided a detailed description.

Ditopella asclepiadea Bonar, Mycologia **57**: 382. 1965, does not appear from the description to belong to this genus. The small perithecia have a slightly papillate beak and are scattered or grouped in cinereous spots which bear a reticulum of dark brown hyphae. Asci were described as bitunicate, 32 spored, 80-120 × 30-40 μm, nonparaphysate, and ascospores as 15-21 × 7-8 μm, ellipsoid, hyaline, one celled. This description suggests either a member of the Dothioraceae, perhaps a species of *Delphinella* if the ascospores become septate, or the genus *Plurisperma* Sivanesan (**1970**) (*Muellerella* according to von Arx and Müller, **1975**).

41. **Phragmoporthe** Petrak, Ann. Mycol. **32**: 354. 1934.

Stromatic tissues pseudoparenchymatous, reduced to ectostromatic disc surrounding and united with beak tissues, extending into periderm as flaring clypeus; perithecia immersed separately, upright, beaks broad. Asci ellipsoid. Ascospores hyaline, ellipsoid, usually slightly curved, several septate.

In branches and petioles of deciduous trees, also reported in Gymnosperms.

Type species: *Phragmoporthe conformis* (Berk. & Br.) Petrak.

Reid and Booth (**1967**) described and illustrated *P. conformis* from European specimens and provided the synonymy given here. The North American collections cited below agree with European ones. *Phragmoporthe conformis* and *Ditopella ditopa* are often found on the same twigs in Europe and are quite similar in all respects except for the octosporous or polysporous asci and ascospore septation.

1. **Phragmoporthe conformis** (Berk. & Br.) Petrak, Ann. Mycol. **39**: 285. 1941. FIG. 114

Sphaeria conformis Berk. & Broome, Ann. Mag. Nat. Hist. 2, **9**: 325. 1852.

Metasphaeria conformis (Berk. & Br.) Sacc., Atti Inst. Venet. Sci. Lett. Art. 6, **2**: 440. 1883-84.

Sphaeria ditopa forma *octospora* Cooke, Handbook of British Fungi **2**: 888. 1871.

Valsa (Calospora) alnicola Cooke & Massee, Grevillea **16**: 47. 1887.

Calospora alnicola (Cooke & Massee) Sacc., Syll. Fung. **9**: 872. 1891.

Massarina alnicola (Cooke & Massee) Berl., Icon. Fung. **1**: 118. 1892.

Phragmoporthe alnicola (Cooke & Massee) Petrak, Ann. Mycol. **38**: 209. 1940.

Massarina ploettneriana P. Henn., Abk. Bot. Ver. Prov. Brandenburg **41**: 98. 1899.

Phragmoporthe ploettneriana (P. Henn.) Petrak, Ann. Mycol. **32**: 354. 1934.

Sphaerulina alni A.L. Smith, Trans. Brit. Mycol. Soc. **6**: 151. 1919.

Ectostromatic disc 500-550 μm diam, perithecia 800-950 μm diam, 400-500 μm high, beak 200-380 μm long, 100-200 μm wide. Asci (42-) 63-75 × 14-21 μm, eight spored. Ascospores (14-)17-25 × 5.5-9 μm, (two) three (four) septate.

In petioles and twigs of *Alnus* spp., Europe, North America.

Material examined: *Europe:* Rehm: Ascom. 1139; Roumeguere Fungi sel. exs. 7145 (NY as *Metasphaeria conformis*, bearing both *P. conformis* and *Ditopella ditopa*). *North America: British Columbia:* Barr 467 (UBC); 5581 (MASS).

Funk (**1975**) has described a second species, *Phragmoporthe pseudotsugae*, on branches of *Pseudotsuga menziesii* (Mirb.) Franco in British Columbia. This species has smaller perithecia than in *P. conformis*, cylindroid asci 95-110 × 8-9 μm, and three-septate ascospores 14-17 × 5-6.5 μm. The description and figures suggest *Griphosphaeria corticola* (Fckl.) v. Höhnel, which does occur in coniferous as well as deciduous tree branches.

III-2b. **Pseudovalsaceae** subfamily **Pseudovalsoideae** tribe **Pseudovalseae**

Perithecia grouped, stromatic tissues compact and forming stromatic disc around beaks, ±diffuse at sides, base often delimited by blackened marginal zones.

Key to Genera of Tribe Pseudovalseae

1. Ascospores broadly ellipsoid, consistently one septate, appendaged; stromatic tissues closely surrounding one or few perithecia 2
1. Ascospores ellipsoid or elongate or fusoid, one to several septate, not appendaged; stromatic tissues forming isolated pulvinate or lens-shaped erumpent structures containing numerous perithecia .... 44. *Pseudovalsa*
    2. Ascospores light to dark brown, appendages narrow, short or elongate............ .... 42. *Chapeckia*
    2. Ascospores hyaline or yellowish, appendages elongate, straplike .... 43. *Macrodiaporthe*

Titania Berl. (Icon. Fung. **1**: 49. 1894), with the type and only species *T. berkeleyi* Berl., was based on a single collection, *Diatrype titan* Berk. & Rav., Grevillea **4**: 97. 1876. It apparently belongs in this subfamily, related to or merely a variant of *Pseudovalsa* species with narrow ascospores. The description and illustrations provided by Wehmeyer (**1941b**) show perithecia ±grouped in dark stromatic tissues, beaks erumpent, and the one ascospore in an ascus dark brown, cylindroid fusoid, five to eleven septate, and 76-133 × 15-23 μm. As Wehmeyer suggested, the one-spored asci upon which the genus was based may be an abnormal development of some other fungus. Additional collections should be made of *Carpinus* from South Carolina, in effort to rediscover the fungus and to determine its true relationships.

Pseudocryptosporella Reid & Booth (Canad. J. Bot. **47**: 1058. 1969), with the type species *P. polylepidis* (Müller) Reid & Booth, from South America, was separated from *Cryptosporella* on the bases of polystichous insertion of perithecia in a pulvinate ±pseudoparenchymatous stroma, and separately erumpent somewhat broad beaks with sulcate apices. The authors compared it to *Sillia* in stroma and perithecial characteristics. It appears to be an amerosporous genus of tribe Pseudovalseae.

Savulescua Petrak (Ac. Rep. Pop. Rom. p. 591. 1959) is typified by *S. insignis* Petrak, which forms tumor-like swellings in living branches of a melastomataceous plant in Puerto Rico. The stromatic tissues are pseudoparenchymatous, asci are four spored, and the hyaline ascospores are broadly ellipsoid and one septate.

Pseudothis Theissen & Sydow (Ann. Mycol. **12**: 274. 1914), with *P. coccodes* (Lev.) Theissen & Sydow, is parasitic in leaves of leguminous plants in particular in tropical regions. The pseudoparenchymatous stromata become erumpent from the epidermis as pulvinate structures. Ascospores are reddish brown, almost truncate at the tips, with a submedian septum.

42. **Chapeckia** Barr, genus novus

Melanconis subgenus *Pseudomelanconis* section *Nigrosporae* Wehmeyer, Revision of Melanconis ... p. 65. 1941.

Stromata pseudoparenchymata, perithecia pauca, globosa, ostiola lata et brevierumpente. Asci unitunicati, tenui paraphysati. Ascosporae brunneae, ellipsoidae vel ovoideae, uniseptate, parietes scabridi, extrema mucronata. Species typica: *Diatrype nigrospora* Peck.

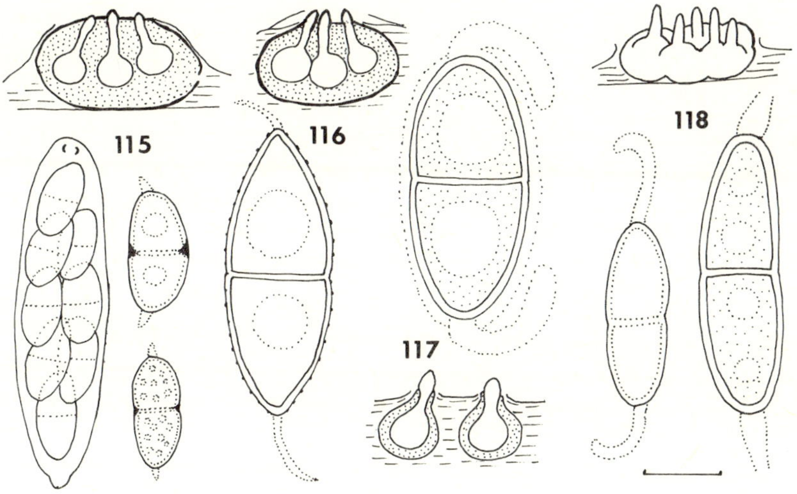

FIGS. 115-118. FIGS. 115, 116. Species of *Chapeckia*. 115. *C. nigrospora,* outline of perithecia in stroma, ascus, ascospores. 116. *C. ribesia,* outline of perithecia in stroma, ascospore. FIGS. 117, 118. Species of *Macrodiaporthe*. 117. *M. occulta,* section of perithecia in stroma, ascospore. 118. *M. everhartii,* habit of perithecia in stroma, ascospores. Scale = 1 mm or 10 μm.

Stromatic tissues pseudoparenchymatous, blackened externally, brown internally, forming isolate eutypoid stromata; perithecia upright, few in stroma, beaks broad, barely erumpent or elongate beyond stroma surface. Asci ellipsoid. Ascospores light to dark brown, ellipsoid, one septate, wall surrounded by narrow gel coating which dries down as irregular roughenings, appendages narrow, short or elongate.

In branches of deciduous trees and shrubs.

Type species: *Diatrype nigrospora* Peck.

1. **Chapeckia nigrospora** (Peck) Barr, comb nov. FIG. 115

Diatrype nigrospora Peck, New York State Museum Rep. **33**: 33. (for 1879) "1880" 1883.

Valsaria nigrospora (Peck) Berl. & Vogl., Syll. Fung. Addit. p. 129. 1886.

Melanconiella nigrospora (Peck) Dearn. & House, New York State Mus. Bull. **266**: 81. 1925.

Melanconis nigrospora (Peck) Wehm., Mycologia **32**: 327. 1940.

Melanconis meschuttii (Ellis & Everh.), Bull. Torrey Bot. Club **10**: 117. 1883.

Melanconiella meschuttii (Ell. & Everh.) Berl. & Vogl., Syll. Fung. Addit. p. 129. 1886.

Stromata immersed in and erumpent from periderm, 700-2000 μm diam, rounded or elliptic, dark brown, blackened externally, composed of compact pseudoparenchymatous tissues; perithecia few (three to four or up to nine) in a stroma, deeply immersed, 330-440 μm diam, beaks short and wide, 275-550 μm long, 165-190 μm wide at flared upper portion; peridium narrow, 20-25 μm wide. Asci 50-90 × 11-14(-16) μm, oblong ellipsoid. Ascospores 13-18 × 6-8 μm, appendages 2-3 × 1 μm.

In branches of *Betula lutea* Michx. f., North America.

Material examined: *Vermont:* Barr 4213 (MASS); *Massachusetts:* Barr 2789, 2953a, 2954, 2974, 5213, 5322 (MASS). *New York:* Quaker Street (Albany), Jun 1879, C.H. Peck (type of *Diatrype nigrospora*, NYS).

Valsaria niesslii (Wint.) Sacc. also occurs on *Betula* branches and has many similarities with *Chapeckia nigrospora*. However, in *V. niesslii* the stromatic tissues are delimited by a black line, perithecia are not deeply immersed in stroma, asci are cylindroid and stipitate, and ascospores are slightly smaller (9-12 × 3.5-4.5 μm) with short rounded appendages and apparent germ pore at each end.

2. **Chapeckia ribesia** (Sacc. & Scalia) Barr, comb. nov. FIG. 116

Pseudovalsa ribesia Sacc. & Scalia in Saccardo, Peck & Trelease, Harriman Alaska Exped. **5**: 29. 1904.

Aglaospora ribesia (Sacc. & Scalia) Farlow, Bibliog. Index, p. 166. 1905.

Stromata 1-2 mm diam, perithecia 4-7 in a stroma, 200-350 μm diam, beaks short erumpent beyond stroma surface. Asci up to 260 μm long, 22-30 μm wide. Ascospores 37.5-46 × 5-16.5 μm, hyaline to light brown, wall roughened, appendages narrow, 6-17.5 μm long.

In branches of *Ribes* spp., western North America.

Material examined: *Oregon:* Swim, 19 Jul 1927, L.N. Goodding; Wilson River, 24 May 1928; Larch Mt., Muttunah Co., 30 May 1928 (all BPI).

Chapeckia ribesia is larger throughout than *C. nigrospora*, but it is identical in stroma and perithecial configuration as well as in the shape, pigmentation, and appendages of ascospores.

The collections cited above were previously identified as *Neokiessleria ribis* (P. Henn. & Ploettn.) Petrak, but that fungus has circinate perithecia beneath a valsoid stromatic disc and is quite different from the present one. The original description and illustrations of *Pseudovalsa ribesia* Sacc. & Scalia (in Saccardo, Peck, and Trelease, **1904**) agree reasonably well with the Oregon material. The asci were described as 110-130 × 24-26 μm, and the ascospores were said to become three septate, and were slightly wider, 38-41 16-20 μm. A gelatinous coating was described; such a coating apparently was present in the Oregon collections, for the wall roughening is of a type consistent with dried-down gel coating. The type collection no larger bears any material of the pyrenomycete, according to Wehmeyer (**1941**), who considered *P. ribesia* to be a doubtful species on that account.

43. **Macrodiaporthe** Petrak, Ann. Mycol. **17**: 94. 1919.

Calospora Nits. ex Fuckel, Symb. Mycol. p. 190. 1870, nom. nud., non Saccardo.

Stromatic tissues pseudoparenchymatous, forming a firm carbonaceous layer around individual perithecia or small groups of few perithecia; perithecia upright or oblique in stroma, beaks broad, central or eccentric at times, erumpent a short distance beyond surface of stroma. Asci broadly ellipsoid. Ascospores yellowish, broadly ellipsoid, one septate, appendages elongate, straplike, surrounded by narrow gel coating.

Type species: *Macrodiaporthe occulta* (Nits.) Petrak.

Petrak (**1971c**) recognized that his genus was based on the same fungus as was *Calospora* Nits., and reiterated that *Calospora* was indeed distinct from *Melanconis*. Petrak's name *Macrodiaporthe* must be utilized for *Calospora occulta* Nits. in Fuckel. Holm (**1975b**) observed that *Calospora* Nits. ex Fuckel (**1870**) was a nomen nudum, and that *Calospora* Nits. ex Niessl (**1875**) was either a nomen nudum or a non validly published name, depending on one's interpretation of Niessl's intention. *Calospora* Saccardo (**1883**) is discussed later under the synonyms of *Prosthecium*.

1. **Macrodiaporthe occulta** (Nits.) Petrak, Ann. Mycol. **17**: 94. 11919. FIG. 117

Calospora occulta Nits. in Fuckel, Symb. Mycol. p. 190. 1870.

Melanconis occulta (Nits.) Sacc., Syll. Fung. **1**: 605. 1882.

Melanconidium occultum (Nits.) Kuntze, Rev. Gen. Pl. **3**(2): 493. 1898.

Diaporthe abnormis v. Höhnel, Sitzungsber. Kaiserl. Akad. Wiss., Math.-Naturwiss. Cl., Abt. 1, **126**: 386. 1917.

Stromata up to 700 μm diam, closely surrounding one perithecium, blackened, 65-80 μm wide, of pseudoparenchymatous cells; perithecia 350-550 μm diam, peridium 26-40 μm wide, brown externally, beaks relatively broad, tips erumpent beyond stromatic tissues, reddish brown. Asci 100-145 × 25-40 μm. Ascospores 36-51 × 12-18 μm, with narrow gel coating, appendages 15-22 μm long, 3-3.5 μm wide.

In branches of *Populus* spp., Europe, North America.

Material examined: *New York:* New York Fungi 171; Ellis & Ev. N.A.F. 3030, 3125 (all MASS).

Macrodiaporthe occulta has ascospores similar in shape and appendages to those of *Pseudovalsella;* the circinate perithecia with beaks converging through the stromatic disc of the latter genus place it in the Melanconidaceae. Another fungus approaching *M. occulta* is *Melanconis everhartii.* The stromatic tissues closely surrounding one or few perithecia and the ascospore shape and appendages are similar in both species, so that the transfer of *M. everhartii* to *Macrodiaporthe* is proposed.

2. **Macrodiaporthe everhartii** (Ellis) Barr, comb. nov. FIG. 118

Melanconis everhartii Ellis, Bull. Torrey Bot. Club **10**: 117. 1883.

Melanconidium everhartii (Ellis) Kuntze, Rev. Gen. Pl. **3**(2): 493. 1898.

Melanconis dasycarpa Ellis & Kellerman, J. Mycol. **2**: 3. 1886.

Melanconidium dasycarpum (Ellis & Kellerm.) Kuntze, Rev. Gen. Pl. **3**(2): 493. 1898.

Valsa magnispora Ellis & Everh., J. Mycol. **3**: 42. 1887.

Diaporthe magnispora (Ellis & Everh.) Sacc., Syll. Fung. **9**: 707. 1891.

Cryptodiaporthe magnispora (Ellis & Everh.) Wehm., The Genus *Diaporthe*, p. 204. 1933.

Stromatic tissues brown, blackened above; perithecia 300-400 μm diam, beaks short to long erumpent, thickened with continuation of stromatic tissues through much of length. Asci 88-130 × 16.5-22 μm. Ascospores 24-39 × 7.5-10.5 μm, hyaline, one septate, appendages up to 26 μm long in fresh material.

In branches of *Acer* spp., North America.

Material examined: *Massachusetts:* Amherst, P.J. Anderson; Barr 5338, 5339, 5651b (MASS); *New York:* Forest Home, Ithaca, 23 Jun 1925, J.H. Miller (GA 6326 as *Cryptodiaporthe magnispora*); *Michigan:* Barr 1320 (MASS); *New Jersey:* Ellis N.A.F. 756 (MASS); *Pennsylvania:* Ellis N.A.F. 1565 (MASS); *Kansas:* Ellis N.A.F. 1561 (MASS).

The description and illustration of *C. magnispora* in Wehmeyer (**1933**) was strongly suggestive of this species, and the New York collection determined by J.H. Miller is certainly not a *Cryptodiaporthe* and is identical with *M. everhartii.*

44. **Pseudovalsa** Ces. & de Not., Comment. Soc. Critt. Ital. **1**: 206. 1863.

Melanconis subgen. *Pseudomelanconis* sect. *Modoniae* Wehm., A revision of *Melanconis*, p. 63. 1941.

Stromatic tissues pseudoparenchymatous, forming a compact isolated ectostroma and variably well or poorly developed dark entostromatic tissues, ectostromatic disc composed of small-celled textura prismatica, entostroma larger celled, more irregular, or interwoven, often with cells of host periderm included; perithecia immersed beneath ectostromatic disc, upright, beaks central, broad, at times somewhat converging and erumpent at surface of disc or extending beyond surface, aspect sulcate at times. Asci ellipsoid to cylindroid. Ascospores ellipsoid, fusoid, or elongate, hyaline, soon brown, at times remaining hyaline, end cells often lighter brown or hyaline, straight or curved, one to several septate, often distoseptate, not appendaged.

Conidial state: *Coryneum* species where known. Acervuli pseudoparenchymatous, conidiophores septate, conidiogenous cells annellidic; conidia ellipsoid, fusoid, or elongate, brown, apical cell often hyaline at tip, straight or curved, several distoseptate.

In branches of deciduous trees.

Type species: *P. lanciformis* (Fr.) Ces. & de Not.

Pseudovalsa in the present sense includes species whose ascospores are permanently one-septate as well as species with several-septate ascospores. The amount of entostromatic tissues formed varies among species. Ectostromatic tissues are compact, and usually form a disc through which the beaks barely emerge. Occasionally the beaks become elongated beyond the surface of the disc. *Coryneum* states are known (or associated) with most of the species. The ascospores may have thickened walls and septa and homogeneous cell contents, or they may be distoseptate with the cell lumina reduced in size. The species included in *Pseudovalsa* fall into two groups, based upon the shapes of ascospores. *Pseudovalsa lanciformis* and related species have broad ascospores whose length: width ratio is 2-3:1, while the other species have narrower ascospores, with a length: width ratio of 4 or more:1.

Cesati and de Notaris (**1863**) noted the similarity of ascospores of *Pseudovalsa lanciformis* and *Aglaospora profusa* (Fr.) de Not., and Rehm (**1906**) and von Höhnel (**1906**) believed that only one genus should be recognized. In the compact, isolated stromata, deeply immersed perithecia, and broad, often sulcate apices of beaks, the two are superficially similar. The ascospores are also superficially similar, but in *A. profusa* the pigmentation is in the lens-shaped cells rather than in the thickened walls. Centrum characteristics are of basic importance to separate the two: the centrum of *P. lanciformis* is diaporthoid and "paraphyses" are the broad cellular remnants of centrum pseudoparenchyma, while in *A. profusa* the centrum (locule) contains narrow pseudoparaphyses in a gelatinous matrix. In surface view, the peridium cells of perithecia of *P. lanciformis* form the diaporthoid textura epidermoidea but this is not apparent in *A. profusa*. *Aglaospora* is a member of the Loculoascomycetes belonging in the Massariaceae of the Melanommatales (Barr, **1976a**).

1. **Pseudovalsa lanciformis** (Fr.) Ces. & de Not., Comment. Soc. critt. Ital. **1**: 206. 1863. FIG. 119

Sphaeria lanciformis Fries, Syst. Mycol. **2**: 362. 1823.

Diatrype lanciformis (Fr.) Fries, Summa Veget. Scand. p. 385. 1849.

Melanconis lanciformis (Fr.) Tul., Ann. Sci. Nat. Ser. 4, **5**: 110. 1856.

Sphaeria melasperma Fries, Syst. Mycol. **2**: 389. 1823.

Sphaeria favacea Tul., Ann. Sci. Nat. Ser. 3, **15**: 377. 1849, non Fries, non Desm.

Melanconis elliptica Peck, Ann. Rep. New York State Mus. **25**: 102. 1873.

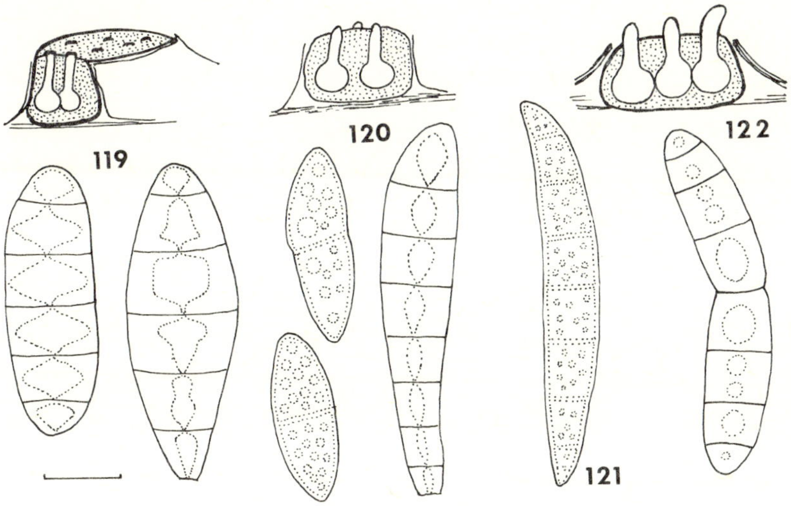

FIGS. 119-122. Species of *Pseudovalsa.* 119. *P. lanciformis,* outline of perithecia in stroma, ascospore, conidium. 120. *P. modonia,* outline of perithecia in stroma, ascospores, conidium. 121. *P. longipes,* ascospore. 122. *P. spiniferum,* outline of perithecia in stroma, ascospore. Scale =11 mm or 10 μm.

Diatrype elliptica (Peck) Peck, Ann. Rep. New York State Mus. **28**: 87. 1876.

Pseudovalsa lanciformis var. *elliptica* (Peck) Sacc., Syll. Fung. **2**: 135. 1883.

Pseudovalsa malbrancheana Sacc., Michelia **1**: 509. 1881.

Aglaospora malbrancheana (Sacc.) Sacc., Syll. Fung. **2**: 134. 1883.

Pseudovalsa betulae Schum. ex Schroet. in Cohn, Kryptogamenfl. Schlesiens **3**(2): 443. 1897.

Aglaospora betulae (Schum. ex Schroet.) Kuntze, Rev. Gen. Pl. **3**(2): 441. 1898.

Stromata rounded to elliptic, horizontal on branch, up to 1.5-2.5 × 0.5-1 mm, erumpent disc olive blackish, interior lighter grayish or brownish black; perithecia seated in base of stroma, 385-440 μm diam, 300-330 μm high, wall thin; beaks broad, dark externally in stroma, apices broad, black and shining, small and indistinct or coarse and appearing sulcate in surface view. Asci 90-170 × 20-30 μm. Ascospores (20-)30-45(-50) × 10-15 μm, (three) five septate, not constricted at septa, walls thickened and brown, end cells lighter brown to hyaline.

Conidial state: *Coryneum brachyurum* Link: Conidia 35-50(-60) × (13-)15-18(-21) μm, three, four, or five (six) distoseptate.

On branches of *Betula* spp., Europe, North America.

Material examined: *Massachusetts:* Pelham, 6 Apr 1919, P.J. Anderson (MASS); *New York:* Karner, Apr 1871, C.H. Peck (lectotype of *Melanconis elliptica* sec. Wehmeyer 1941b), Nov 1871 (NYS); *Georgia:* GA 6763, 6764, 6765 (GA).

Wehmeyer (**1941b**) included in the species concept *Pseudovalsa malbrancheana* Sacc., on *Prunus spinosa*, with ascospores 30-47.5 × 12.5-17.5 μm, in four- or eight-spored asci. Gilman, Tiffany and Lewis (**1959**) reported *P. lanciformis* on *Prunus* from Iowa.

Wehmeyer (**1941b**) suggested that *Pseudovalsa umbonata* (Tul.) Sacc. (Fung. Venet. **4**: 20. 1875) also could be a substrate variant of *P. lanciformis*. *Pseudovalsa umbonata* occurs on species of *Quercus* in Europe. The stromata are less elliptic than normally in *P. lanciformis* and the ascospores measure 35-56 × 13-18 μm. The conidial state is *Coryneum depressum* Schmidt ex Steudel, with the conidia measuring 40-50(-60) × (17-)19-22(-25) μm, broader than those of *C. brachyurum*. *Pseudovalsa umbonata* is retained as a closely related but distinct species from *P. lanciformis*. Gilman, Tiffany and Lewis (**1959**) reported it from Iowa.

2. **Pseudovalsa modonia** (Tul.) v. Höhnel, Ann. Mycol. **16**: 125. 1918.
FIG. 120

Melanconis modonia Tulasne, Ann. Sci. Nat. ser. 4, **5**: 111. 1856.

Melanconidium modonium (Tul.) O. Kuntze, Rev. Gen. Pl. **3**(2): 493. 1898.

Pseudovalsella modonia (Tul.) Kobayashi, Bull. Gov. Forest Exp. Sta. Tokyo **226**: 35. 1970.

Sphaeria biconica Currey, Trans. Linn. Soc. London, **22**: 279. 1858.

Stromatic tissues yellowish brown to dark brown, forming an erumpent ectostromatic disc; perithecia deeply immersed, 220-700 μm diam, beaks reaching surface of disc. Asci 88-150 × 15-21 μm. Ascospores 26-38.5 × 6.5-12 μm, hyaline, finally brown after full maturity, one (three) septate, wall and septum thick, at times surrounded by narrow gel coating.

Conidial state: *Coryneum* spp. The European *P. modonia* is associated with *Coryneum modonium* (Sacc.) Griff. & Maubl.; conidia are five to eight distoseptate, (44-)50-71(-75) × 14-19(-22) μm. In North American material, the conidial state is instead referable to *C. castaneicola* Berk. & Curt. in Berk.; conidia are narrower, 45-70(-80) × (9-)10-13 (-15) μm, five to seven distoseptate. The two species were separated by Sutton (**1975**). The conidia of Japanese specimens described by Kobayashi (**1970**) are narrower than the European and also appear to be *C. castaneicola.*

On branches of *Castanea* spp., Europe, North America, Asia.

Material examined: *Massachusetts:* Barr 2724a, 4835b, 5192, 5361 (MASS); *New York:* New York Fungi 336 (MASS); *New Jersey*: Ellis & Everhart N.A.F. 1564 (MASS).

The ascospores of *P. modonia* rarely form more than one septum, and Wehmeyer (**1941b**) preferred to retain the species in *Melanconis*. The species deviates from those of *Melanconis* by the upright perithecia beneath a well developed brown ectostromatic disc, surrounded by brown entostroma, and by the conidial state. *Pseudovalsella* includes species with similar ascospores, except that there they are appendaged. The stromatic tissues in species of *Pseudovalsella* are less dense, lighter colored, and are often delimited by a blackened marginal line. The conidial state asscociated with species of *Pseudovalsella* is borne in stromatic cavities rather than on acervuli.

Pseudovalsella tetraspora Kobayashi (Bull. Gov. Forest Exp. Sta. Tokyo **226**: 37. 1970) on *Betula* branches, is related to *Pseudovalsa modonia* in ascospore size and septation. The ascospores of *P. tetraspora* are consistently brown. The conidial state, *Coryneum pedunculatum* Kobayashi (non *C. pedunculatum* (Ell. & Ev.) Hughes, **1955**) is closely related to *C. brachyurum* according to Sutton (**1975**).

Species with narrowly ellipsoid or fusoid, elongate ascospores are exemplified by *Pseudovalsa longipes.*

3. **Pseudovalsa longipes** (Tul.) Sacc., Fung. Venet. **4**: 20. 1875. FIG. 121

Melanconis longipes Tul., Ann. Sci. Nat. ser. 4, **5**: 111. 1856.

Calospora longipes (Tul.) Berlese, Icon. Fung. **1**: 117. 1894.

Sphaeria quercina Currey, Trans. Linn. Soc. London **22**: 272. 1858, non Persoon, non auct.

Aglaospora quercina ("Fries") Kuntze, Rev. Gen. Pl. **3**(2): 441. 1898.

Sphaeria arcuata Currey, Trans. Linn. Soc. London **22**: 281. 1858.

Melanconis sigmoidea Cooke & Ellis, Grevillea **6**: 92. 1878.

Pseudovalsa sigmoidea (Cke. & Ell.) Sacc., Syll. Fung. **2**: 136. 1883.

Aglaospora sigmoidea (Cke. & Ell.) Kuntze, Rev. Gen. Pl. **3**(2): 441. 1898.

Calospora ambigua Pass., Diagn. Fung. Nuov. **4**: 31. 1890.

Khekia ambigua (Pass.) Petrak, Hedwigia **62**: 284. 1921, nomen, non spec.

Stromata dark brown, blackened externally; perithecia 300-500 μm diam, beaks up to 450 μm long. Asci 75-125(-180) × 17-25 μm. Ascospores (27-)44-60(-80) × 6-8(-11) μm, hyaline or yellowish, becoming brown in age, usually curved, elongate fusoid, three, four, five, six to ten septate.

Conidial state: *Coryneum umbonatum* Nees ex Steudel: acervuli formed on exposed surfaces of young stroma; conidia (40-)50-60(-117) × (10-)13-16(-18) μm, five to seven (nine) distoseptate, brown, apical cell darker than other cells (Sutton, **1975**).

On branches of *Quercus* spp., Europe, North America.

Material examined: *Ontario:* Ellis & Everhart N.A.F. 2758 (MASS); *Rhode Island:* Barr 4760 (MASS); *New Jersey:* Ellis & Everhart N.A.F. 1563, Fungi Col. 1423, 1427 (MASS); *Pennsylvania:* E. Bartholomew Fungi Col. 4251 (MASS); *Georgia:* GA 6767, 6768 (GA); *Louisiana*: Barr 6332b (MASS).

Melogramma spiniferum (Wallr.) de Not. should also be considered as a species of *Pseudovalsa. Melogramma* Fries (**1849**) emend. Tul. & Tul. (**1863**) has for type species *M. bulliardi* Tul. (*M. vagans* de Not.), which differs markedly from *M. spiniferum.* Thus typified, *Melogramma* is excluded from my concept of the Diaporthales and is a member of the Phyllachorales (sensu Barr, **1976b**). Doguet (**1961**) described *Melogramma spiniferum* in detail. He paid special attention to the ascus apex and spore ejection. This he described as the "Jack in box" type, and he characterized the asci as bitunicate with a diaporthoid apex. In my opinion, the asci are unitunicate. Stromatic and perithecial configuration are those of *Pseudovalsa*, although the beaks are elongate and extend beyond the stromatic surface, in contrast to those of other species of *Pseudovalsa.* Ascospores are distoseptate, and the cell lumina are reduced in size. No conidial state is known for this species.

4. **Pseudovalsa spinifera** (Wallr.) Barr, comb. nov. FIG. 122

Sphaeria spinifera Wallr., Crypt. Flora **2**: 846, no. 4073. 1833.

Melogramma spiniferum (Wallr.) de Not., Sfer. Ital. p. 53. 1863.

Diatrype podoides Fries, Summa Veget. Scand. p. 385. 1849.

Melogramma podoides (Fr.) Auersw. in Niessl, Crypt. Fl. Mähren, p. 109. Verh. Naturf. Vereins Brünn **3**: 109. 1864.

Melogramma asperum Ces. & de Not., Comment. Soc. Crittog. Ital. **1**(4): 30. 1863.

Stromata 2-4 mm diam, interior brown, pseudoparenchymatous; perithecia 330-495 μm diam, immersed in base of stroma; beaks broad, extending up to 1 mm beyond stroma surface, apices reddish, pigment leaching in KOH. Asci 85-120 × 12-15 μm, four or eight spored. Ascospores (38-)48-70 × (5-)6-8 μm, hyaline at first, soon brown with hyaline or light brown end cells, elongate, often slightly curved, (one, three) five to seven septate.

On branches of *Fagus*, Europe; *Betula*, North America.

Material examined: *Ontario:* Ellis & Everhart, N.A.F. 3032 (MASS); *Michigan:* Barr 5384 (MASS).

A collection on *Sambucus callicarpa* from California (Barr 5950a, MASS) also apears to be a species of *Pseudovalsa*. The asci are 140-180 × 15-18 μm, with the apical annulus clearly visible in water mounts and chitinoid in nigrosin. The distoseptate ascospores with (one) three, four or five septa measure (25-)35-45 × 8-11 μm. An associated conidial state produces conidia which are brown, seven distoseptate, and 36-45 × 10-12 μm. These conidia are similar to but shorter than those of *Coryneum japonicum* (Sacc.) Sutton on *Quercus* (Sutton, **1975**). The fungus certainly differs from *Pseudovalsa sambucina* (Peck) Sacc., a loculoascomycete belonging in *Thyridaria* (Wehmeyer, **1941a**). Additional collections are needed to clarify the range of variations in this taxon.

IV. Family **Melanconidaceae** Winter in Rabenh., Kryptogamen-Flora... 1(2): 764. 1887, Barr, emend.

Perithecia oblique or horizontal, beaks oblique or lateral, erumpent separately or converging and erumpent through stromatic disc. Type genus: *Melanconis* Tulasne & Tulasne. (2 subfamilies).

Key to Subfamilies and Tribes of Melanconidaceae

1. Perithecia grouped or single, beaks separate or converging .... Subfamily Melanconidoideae . 2
1. Perithecia grouped, beaks fused together to form pseudostromatic disc .... Subfamily Massariovalsoideae . 3
 2. Perithecia single and beaks separate, stromatic tissues prosenchymatous, forming clypeus at surface of substrate .... Tribe Plagiophialeae, p. 176
 2. Perithecia grouped and beaks convergent, stromatic tissues prosenchymatous, forming disc at surface of substrate .... Tribe Melanconideae, p. 179
3. Prosenchymatous tissues slight, without blackened marginal zones.. Tribe Massariovalseae, p. 189
3. Prosenchymatous tissues well developed, delimited by blackened lateral and ventral zones .... Tribe Hercosporeae, p. 193

Subfamily IV-1. **Melanconidoideae**

Perithecia grouped or single, beaks converging or erumpent separately, stromatic tissues prosenchymatous, as blackened zone or clypeus or stromatic disc at substrate surface, at times forming blackened marginal zones in substrate. (2 tribes).

IV-1a. Tribe **Plagiophialeae** Barr, tribus nov.

Perithecia singulara, rostris erumpentis, lateralis, stromata contexta prosenchymata, saepe instar clypei; in foliis dicotyledonibus arboribus et frutibus et herbarum. Genus typicus: *Plagiophiale* Petrak.

One genus in temperate North America: 45. *Plagiophiale.*

Perithecia separate or few grouped, stromatic tissues prosenchymatous, forming slight or well developed clypeus, in parasitic members producing leaf spots permeated by hyphae.

Plagiostigme Sydow (Ann. Mycol. **23**: 341. 1925) is typified by *P. couraliae* Sydow, a leaf-spotting parasite on a member of the Bignoniaceae in tropical America. Prosenchymatous hyphae form a darkened pseudostroma between cells of the leaf tissues in which the oblique perithecia with short lateral beaks are immersed separately. The ascospores are obovoid with a submedian septum and short appendages. *Bioporthe* Petrak (Ann. Mycol. **27**: 24. 1929), with the type species *B. brenesii* Petrak is parasitic also. The oblique perithecia develop in circular groups in pallid spots in leaves of tropical *Tournefortia* (Boraginaceae), and the lateral beaks are erumpent together. Prosenchymatous stromatic tissues are present. The ascospores are cylindroid or narrowed toward the middle; the septum is median. Müller in Müller and von Arx (**1962**) included *Bioporthe* as a synonym of *Plagiostigme*, and **1965** Müller included the type species of *Plagiophiale* in *Plagiostigme*. Petrak (**1965**) disagreed with Müller's concept of *Plagiostigme* and recognized the three genera as separate, based upon habit as well as ascospore shape and position of the septum. I too believe the fungi are not congeneric.

Diatractium Sydow (Ann. Mycol. **18**: 183. 1920) with the type species *D. cordiana* (Ellis & Kelsey) Sydow (Ann. Mycol. **33**: 86. 1935) is another extralimital genus of this tribe. The type species (NY: *Hypospila cordiana* Ellis & Kelsey, on *Cordia collococca*, Bassen, St. Croix, Danish West Indies, 20 Feb 1896, A.E. Ricksecher #25) is parasitic in leaves of *Cordia* spp. in central America. The horizontal perithecia are immersed in leaf spots and open through a blackened clypeus by a short lateral beak. Ascospores are narrow at the median septum, broader above and below, similar to those of *Bioporthe* and of *Pleuroceras* in the Gnomoniaceae. A second species, *D. ingae* (Allesch.) Sydow, is parasitic in leaves of *Inga* spp. *Scolecodothopsis* Stevens (Ill. Biol. Monogr. 8(3): 183. 1923), with the species *S. ingae* Stevens, has filiform ascospores and is a species of *Ophiodothella* (Phyllachorales), rather than *Diatractium* as Müller and von Arx (**1962**) stated.

45. **Plagiophiale** Petrak, Sydowia **9**: 585. 1955.

Stromatic tissues prosenchymatous, slight, forming blackened surface over perithecia and short beak; perithecia oblique or horizontal in substrate, ovoid, beaks oblique or lateral, short and broad. Asci broadly oblong. Ascospores hyaline to yellowish, broadly ellipsoid, one septate, wall surrounded by gel coating, appendages pulvinate.

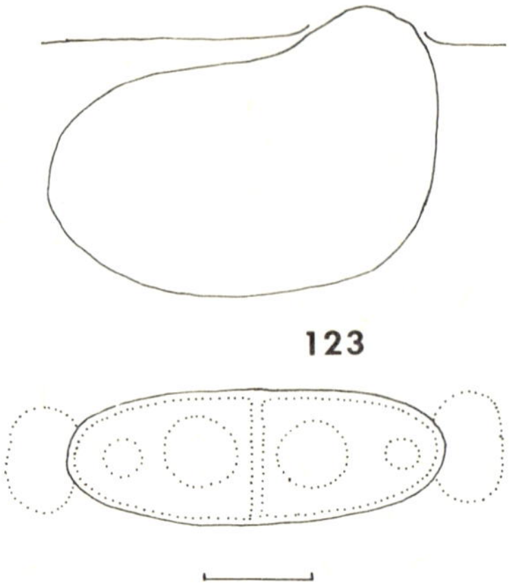

FIG. 123. *Plagiophiale petrakii,* outline of perithecium, ascospore. Scale =100 μm or 10 μm.

Conidial state not known.

In leaves of herbaceous plants.

Type and single species: *Plagiophiale petrakii* (Müller) Petrak.

Petrak originally (**1955c**) identified the fungus which is the type species of *Plagiophiale* as *Sphaerella eucarpa* Karsten. However, Karsten's species belongs in *Wettsteinina* (Müller and von Arx, **1959**; Barr, **1959**; Müller and von Arx, **1962**). Both Barr (**1959**) and Müller and von Arx (**1962**) consigned *Plagiophiale* to synonymy with *Wettsteinina*, believing that Petrak's fungus was identical with *W. eucarpa.* Müller and von Arx noted that Petrak's description was of a sphaeriaceous fungus and that when refound it should be consigned to another genus and given a new name. Müller (**1965**) did this in describing *Plagiostigme petrakii* which is undoubtedly the same fungus as Petrak's *Plagiophiale eucarpa.* I believe that Petrak's (**1965**) arguments are valid and that the genera *Plagiophiale, Plagiostigme,* and *Bioporthe* are separate.

1. **Plagiophiale petrakii** (Müller) Petrak, Sydowia **18**: 387. (1964) 1965. FIG. 123

Plagiophiale eucarpa Petrak, Sydowia **9**: 586. 1955, non *Sphaerella eucarpa* Karsten 1872.

Plagiostigme petrakii Müller, Sydowia **18**: 90 (1964) 1965.

Perithecia 250-280 µm high, 350-450 µm long, including short lateral beak, or beak elongate and distinct, 70-100 µm long and wide. Asci 120-160 × 31-50 µm. Ascospores 32-44 × 12-18.5 µm, gel coating 2-3.5 µm wide while in ascus, appendages 10-14.5 µm wide, 9-11 µm high.

On overwintered leaves of *Polygonum viviparum* L., Europe, North America.

Material examined: *New Hamphire:* Barr 4094 (MASS).

IV-1b. Tribe **Melanconideae**

Perithecia grouped, rarely single in substrate, beaks converging or at times erumpent separately, stromatic tissues prosenchymatous, forming stromatic disc.

Key to Genera to Tribe Melanconideae

1. Ascospores one celled. . . . . 46. *Cryptosporella*
1. Ascospores septate 2
   2. Ascospores muriform. . . . . 50. *Dictyoporthe*
   2. Ascospores transversely septate 3
3. Ascospores one septate, relatively small, up to 27(-34) µm long, appendages short, blunt or pulvinate when formed; conidia typically one celled, hyaline or brown. . . . . 47. *Melanconis*
3. Ascospores one, two or three septate, relatively large, usually 30 µm or longer; conidia typically several septate 4
   4. Ascospore appendages elongate, straplike . . . . 48. *Pseudovalsella*
   4. Ascospore appendages short, blunt or pulvinate. . . . . 49. *Prosthecium*

46. **Cryptosporella** Sacc., Michelia **1**: 30. 1877.

Stromatic tissues prosenchymatous, often light or bright colored and usually forming a small disc through which convergent beaks are erumpent; perithecia oblique or horizontal, grouped, ±circinate, occasionally erumpent separately, beaks relatively broad, short to elongate. Asci ellipsoid. Ascospores hyaline or yellowish, ellipsoid or fusoid, one celled, wall thickened, surrounded by gel coating at times.

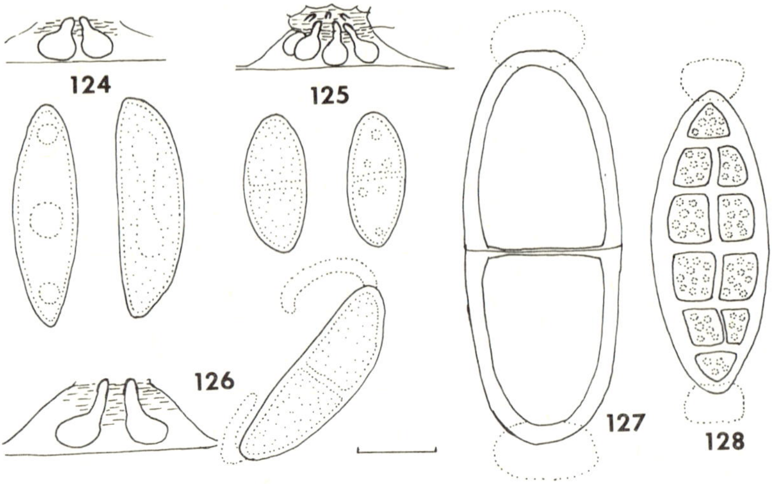

FIGS. 124-128. FIG. 124. *Cryptosporella hypodermia,* habit of perithecia, ascospores. FIG. 125. *Melanconis stilbostoma,* habit of perithecia, ascospores. FIG. 126. *Pseudovalsella thelebola,* habit of perithecia, ascospore. FIG. 127. *Prosthecium acrocystis,* ascospore. FIG. 128. *Dictyoporthe acerophila,* ascospore. Scale =1 mm or 10 μm.

Conidial state where known: *Mastigonetron:* Kleb. (=*Harknessia* Cke., Sutton, **1977**) pycnidia uniloculate or several loculate or locules formed in young stromata; conidiophores annellidic; conidia brown, rarely hyaline, one celled, thick walled, often with apical appendage.

In branches of deciduous trees and shrubs.

Type species: *Cryptosporella hypodermia* (Fr.) Sacc.

Holm (**1975b**) observed that *Wuestneia* Auerswald in Fuckel, Fungi Rhen. n. 587. 1863, was validly published on the label of *W. aurea*; Holm proposed the species as lectotype. *Wuestneia aurea* has been referred to *Cryptosporella.* If this is indeed the case, *Cryptosporella* is a later taxonomic synonym of *Wuestneia.*

Von Arx and Müller (**1954**) erected the family Cryptosporellaceae in the Sphaeriales for *C. hypodermia.* They noted that the family differed from all of the other Sphaeriales by the valsoid aspect of perithecia and ostioles and by the asci, all diaporthoid characters. However, they excluded *Cryptosporella* from the Diaporthales because of the lack of a refractive apical annulus in the ascus. Munk (**1957**) and Reid and Cain (**1960**) did not regard this one feature as sufficient to substantiate a family, since this is a variable character in a number of related genera, and in fact may be be a variable condition in collections of a single species. The North American collections cited below do possess a refractive chitinoid apical annulus when in fresh condition, and the majority of species in the genus show this structure clearly. No data on the conidial state of *C. hypodermia* are available, but Petrak (**1950**) provided information on the *Mastigonetron* conidial states of *C. farinosa* (Ellis) Sacc. and *C. beltsvillensis* Petrak. A collection on *Castanea* (Vermont: Barr 2985, MASS), which appears to be identical with *C. beltsvillensis*, has an associated conidial state which agrees with Petrak's description of *M. affine* (Ell. & Ev.) Petrak (*Harknessia americana* (Mont.) Sutton) except for hyaline rather than dark brown conidia.

1. **Cryptosporella hypodermia** (Fries) Sacc., Syll. Fung. **1**: 466. 1882.
FIG. 124

Sphaeria hypodermia Fries in Kunze and Schwamme, Mycol. Hefte **2**: 49. 1823; Syst. Mycol. **2**: 407. 1823.

Valsa hypodermia (Fr.) Fries, Summa Veget. Scand. 412. 1849.

Cryptospora hypodermia (Fr.) Fuckel, Symb. Mycol. 192. 1870.

Stromatic disc brown, inconspicuous and often obliterated by erumpent beaks; perithecia 350-700 μm diam. Asci 90-120(-200) × (11-)15-20 μm. Ascospores (18-)30-60 × 6.5-10 μm, fusoid.

On branches of *Ulmus*, Europe; *Acer*, decorticated wood, North America.

Material examined: *Vermont:* Barr 4674; *Massachusetts:* Barr 2852 (MASS). Reported from Iowa on *Acer* (Gilman, Tiffany, and Lewis, **1959**).

In *Cryptosporella* a number of variations are found but clarification of species concepts is yet to be completed. From the material at hand, there appear to be at least two groups of species. In *C. hypodermia* and others, e.g. *C. lentaginis* Rehm, the stromatic disc is small, dark or dull brownish or grayish brown, and ascospores are relatively narrowly fusoid and thin walled. *Cryptosporella farinosa* (Ell.) Sacc. exemplifies a group of species where the stromatic disc is light or brightly colored (whitish, yellow, orange, reddish brown, bright brown) and the ascospores are broadly ellipsoid, thick walled and at times surrounded by a gel coating. The number of ascospores per ascus is variable in several of these species of *Cryptosporella: C. paucispora* (Peck) Berl. & Vogl. was described with four-spored asci, but the numbers vary from three to seven; *C. farinosa* and *C. beltsvillensis* Petrak may have from two to eight spores per ascus within a single perithecium.

Cryptosporella acerina Wehm. (in W.B. Cooke, Mycologia **41**: 610. 1949) differs from the present concept of the genus with upright perithecia in brown pseudoparenchymatous stroma. It may represent a temperate-zone species of *Pseudocryptosporella* Reid & Booth, but in *P. polylepidis* (Müller) Reid & Booth the perithecia are embedded at different levels in the stroma and the elongate broad beaks are thickened at the tips.

47. **Melanconis** Tulasne & Tulasne, Sel. Fung. Carp. **2**: 115. 1863.

Melanconiella Sacc., Syll. Fung. **1**: 744. 1882.

Discodiaporthe Petrak, Hedwigia **62**: 293. 1921.

Stromatic tissues prosenchymatous, light colored or brown, gray, or olivaceous, forming a disc through which convergent beaks are erumpent; perithecia oblique to horizontal, ±circinate, beaks lateral, convergent. Asci ellipsoid oblong. Ascospores ellipsoid fusoid or broadly ellipsoid, hyaline to brownish, one septate, appendages short and blunt when present.

Conidial state: *Melanconium* (*Discosporium* when hyaline): as open cavities in upper part of young stroma; conidiophores annellidic; conidia brown, hyaline in some, broadly ellipsoid, ovoid, or subgloboid, one celled.

In branches of deciduous trees.

Type species: *Melanconis stilbostoma* (Fr.) Tul.

Melanconis in the present sense consists of the majority of species which Wehmeyer (**1941b**) placed in subgenus *Eumelanconis*, while those of his subgenera *Massariovalsa* and *Pseudomelanconis* are again separated from *Melanconis*. The genus *Melanconiella*, with type species *M. spodiaea* (Tul.) Sacc., erected to accommodate those species of *Melanconis* in which the ascospores are brown at full maturity, does not differ significantly otherwise and is considered here to be synonymous with *Melanconis*, as Wehmeyer (**1941b**) and Müller and von Arx (**1962**) did. Munk (**1957**), Petrak (**1952a**), Dennis (**1968**) still preferred to separate two genera. *Discodiaporthe* was based on *D. sulphurea* (Fckl.) Petrak for species with hyaline conidia, but Petrak later (**1923**) decided that character was not of generic significance.

The species of *Melanconis* are quite similar to one another, differ in coloration of the ectostromatic disc (white to yellowish, green, olivaceous, gray, or brownish) and in sizes of ascospores and conidia in conjunction with the host plant. Many of the species are found on members of the Corylaceae and Juglandaceae, the others on such host trees as *Tilia* and *Cornus*.

1. **Melanconis stilbostoma** (Fr.) Tul., Ann. Sci. Nat. ser. 4, **5**: 109. 1856. FIG. 125

Sphaeria stilbostoma a papula Fries, Syst. Mycol. **2**: 404. 1823.

Valsa stilbostoma (Fr.) Fries, Summa Veget. Scand. 412. 1849.

Sphaeria socialis Kunze in Wallr., Fl. Crypt. Germ. **2**: 819. 1932.

Sphaeria pulchella Currey, Trans. Linn. Soc. London **22**: 280. 1858.

Valsa bloxami Cooke, Grevillea **14**: 47. 1885.

Diaporthe bloxami (Cke.) Berl. & Vogl., Syll. Fung. Addit. 105. 1886.

Melanconidium bicolor Nees ex Kuntze, Rev. Gen. Pl. **3**(2): 493. 1898.

Ecotostromatic disc creamy to yellowish; perithecia 300-600 μm diam. Asci 85-90 × 15.5-17.5 μm. Ascospores 12-18.5 × 6.5-8(-9) μm, without appendages. Conidia 9-16.5 × 5-7.5 μm, brown.

In branches of *Betula* spp., Europe, North America.

Material examined: *Newfoundland:* Ellis & Everhart Fungi Col. 1323 (MASS); *Massachusetts:* Barr 6250 (MASS); Waverly, 12 Jan 1909, A.T. Speare (GA 6751); *New York:* Labrador Lake, 5 Sep 1925, J.H. Miller (GA 6752).

48. **Pseudovalsella** von Höhnel, Ann. Mycol. **16**: 123. 1918.

Melanconis subgenus *Pseudomelanconis* sect. *Thelebolae* Wehmeyer, A Revision of *Melanconis*... 51. 1941.

Stromatic tissues prosenchymatous, yellowish or brown, forming small isolated areas, often delimited by blackened marginal zone, surrounding small groups of perithecia; perithecia ±circinate, upright, beaks central or eccentric, erumpent to surface or extending beyond disc. Asci broad, ellipsoid oblong. Ascospores hyaline to yellowish, at times brown in age, elongate ellipsoid, one (two or three) septate, appendages elongate, straplike.

Conidial state: where known, as *Hendersonula* Speg. in cavities in stroma; conidia hyaline, soon becoming brown, elongate ellipsoid or fusoid (one) three or more septate. *Hendersoniopsis* v. Höhnel and *Cytosporopsis* v. Höhnel (=*Cytospora*, Sutton, **1977**) for spermatial state.

In branches of deciduous trees.

Type species: *Pseudovalsella thelebola* (Fr.) von Höhnel.

Of the other species in section *Thelebolae* in Wehmeyer's (**1941b**) treatment *M. occulta* is accepted as the type species of *Macrodiaporthe*. *Melanconis helvetica* Rehm in Wehm. perhaps suggests *Macrodiaporthe* also, although ascospores deviate from those of *M. occulta*. *Melanconis corni* Wehm. is apparently a species of *Melanconis*, although its conidial state suggests *Prosthecium*.

1. **Pseudovalsella thelebola** (Fr.) von Höhnel, Sitzungsber. Kaiserl. Akad. Wiss., Math.-Naturwiss. Cl., Abt. 1, **126**: 393. 1917.FIG. 126

Wehmeyer (**1941b**) detailed the synonymy of this fungus as *Melanconis thelebola*; Kobayashi (**1970**) added *Cryptodiaporthe konseiensis* Kob. (Trans. Mycol. Soc. Japan **4**: 8. 1962.)

Stromatic tissues yellowish; perithecia up to 600 μm diam, beaks black and obvious at surface of light colored stroma. Asci 85-120 × 18-28.5 μm. Ascospores 24-35(-42) × 8-10 μm, hyaline, becoming brown in age, one (two, three) septate, appendages elongate, curved.

Conidial state: *Hendersonula thelebola* (Sacc.) Kobayashi.

In branches of *Alnus* spp., Europe, North America, also *Sorbus*, Japan.

Material examined: *New York:* New York Fungi 335; *Montana:* Bartholomew F. Col. 4317 (MASS); *Washington:* E. Bartholomew Fungi Col. 3144 (MASS); *Georgia:* GA 6755, 6756, 6757, 6758 (GA); *California:* Fungi of California, H.E. Parks 4748 (GA).

49. **Prosthecium** Fresenius, Beitr. z. Mykol. **2**: 62. 1852; Wehmeyer **1941b**, pro parte, Barr, emend.

Calospora Sacc., Syll. Fung. **2**: 231. 1883, non Nitschke ex Fuckel, 1870, nec Nitschke ex Niessl, 1875.

Calosporella Schroeter in Cohn, Kryptogamenfl. Schles. **3**(2): 442. 1897.

Phaeodiaporthe Petrak, Ann. Mycol. **17**: 99. 1919. (sub *Pseudodiaporthe* Petrak, Ann. Mycol. **19**: 94. 1921.)

Melanconis subg. *Pseudomelanconis* sect. *Auctae* Wehmeyer, Revision of *Melanconis*... 58. 1941.

Stromatic tissues prosenchymatous, disc erumpent, rounded or elongate, at times with crystalline debris included, at times stromatic tissues extending down around perithecial walls; perithecia oblique in substrate, ±circinate, few to 20 or more, beaks ±lateral, converging and erumpent through disc. Asci broadly oblong, apical annulus refractive or scarcely visible in water mounts, chitinoid or not, paraphyses as broad cellular strips, thin walled and deliquescent. Ascospores hyaline or brown at maturity, ellipsoid or nearly cylindroid, straight or bent, one, two, three (four) septate, not constricted at septa, wall and septa relatively thick to very thick (up to 2.5 μm wide), appendages short, pulvinate or tapered when present, biseriate in the ascus.

Conidial state: *Stilbospora* Pers. ex Mérat where known. Conidia borne in cavities in ectostromatic tissues, on elongate conidiophores; conidia hyaline or brown, ellipsoid, fusoid, elongate, often curved, one celled or one to three septate. Microconidial state formed in cavities in stromatic tissues; conidia hyaline, allantoid, minute.

In woody branches, rarely petioles of leaves of deciduous trees.

Type species: *Prosthecium ellipsosporum* Fresenius.

Prosthecium subgenus *Pseudoprosthecium* Wehmeyer was typified by *P. hapalocystis* (Berk. & Br.) Petrak, but is not closely related to the typical species of *Prosthecium.* In *P. hapalocystis* the minute stromatic disc surrounds fused beaks of a few circinate perithecia; ascospores are constricted at the septa, bear elongate broad appendages, and may be surrounded by a gel coating. *Prosthecium hapalocystis* is more like the species of *Massariovalsa* in stromatic configuration and perithecia. The name *Hapalocystis* is reinstated for the species which have characteristics noted above; the genus is considered further in tribe Massariovalseae.

The generic name *Calospora* Sacc. non Nitschke ex Fuckel (Symb. Mycol. 190. 1870, nom. nud.), nec Nitschke ex Niessl, (Hedwigia **14**: 130. 1875, nom. nud.) was discussed by Holm (**1975b**). *Calospora platanoides* (Pers. ex Fr.) Niessl ex Saccardo, based on Fries Scler. Suec. n. 186, was designated as lectotype species by von Höhnel (**1918**) and was used by Clements and Shear (**1931**). Wehmeyer (**1941b**) used the name *Prosthecium innesii* Currey for this species, declaring that while Fries' interpretation of Persoon's *Sphaeria platanoides* was the *Calo-*

spora platanoides of later authors, Persoon's original species was a different fungus. *Calosporella* Schroeter was proposed to replace *Calospora* Sacc. non Nitschke.

Phaeodiaporthe Petrak (**1919**) (given as *Pseudodiaporthe* by Petrak, **1921**, in error) was typified by *Valsa appendiculata* Otth. This species produces a narrow black marginal zone and one-septate brownish ascospores which have short pulvinate appendages. Because of the one-septate ascospores, Wehmeyer (**1941b**) included *P. appendiculata* under *Melanconis* subgenus *Pseudomelanconis* section *Auctae*. The species of section *Auctae* have more affinity with *Prosthecium* than with *Melanconis* in ascospore shape, appendages, and conidial states; the generic limits of *Prosthecium* are emended here to include species whose ascospores may remain one septate.

Prosthecium ellipsosporum Fres. was described and illustrated by Wehmeyer (**1941b**), who gave details of synonymy. This species, in branches of *Carpinus* in Europe, has not been reported in North America. The ascospores are three septate, brown, 30-44 × 10-14 μm; conidia are three septate, brown, 33-50 × 9-11.5 μm.

Prosthecium platanoides (Fries) Barr, comb. nov. (Basionym: *Sphaeria platanoides* Fries, Scler. suec. 186) is another European species, found in branches of *Acer*. Ascospores are 20-28(-30) × (4-)5-8 μm, hyaline, (one) three (four) septate. No macroconidial state is known with certainty, but a microconidial state is associated with the perithecial state. A specimen described as form *petiolicola* by Petrak (Flora Bohemiae et Moraviae exs. n. 1032) differs from the typical branch-inhabiting form in smaller stromatic discs through which only one or two perithecial beaks are erumpent. *Prosthecium stylosporum* (Ell. & Ev.) Wehm., in branches of *Acer spicatum* Lam. in northeastern North America has brownish, two to three septate ascospores, 22-40 × 9.5-12 μm. The conidial state produces hyaline, three-septate conidia, 33-50 × 5-6.5 μm.

The species which had been in *Melanconis* sect. *Auctae* (Wehmeyer, **1941b**) include: *Prosthecium auctum* (Berk. & Br.) Petrak (Ann. Mycol **21**: 325. 1923) on *Alnus* in Europe. This species has one to three-septate ascospores which become brown at maturity; conidia are hyaline, elongate fusoid, and one septate. **Prosthecium acrocystis** (Peck) Barr, comb. nov. (FIG. 127) (basionym: *Valsa acrocystis* Peck, Ann. Rep. New York State Mus. **33**: 34 (for 1879) "1880" 1883; type — North Greenbush, New York, May, C.H. Peck, NYS) develops in branches of *Betula* species in northeastern North America. The large, one-septate ascospores become brown in age; a presumed conidial state was reported by Wehmeyer (**1941b**). *Melanconis appendiculata* (Otth) Wehm., the third species of section *Auctae*, in Europe occurs on *Acer* species. The combination **Prosthecium appendiculatum** (Otth) Barr, comb. nov. is proposed (basionym: *Diaporthe appendiculata* Otth, Mitth. Nat. Ges. Bern, 100. 1871).

50. **Dictyoporthe** Petrak, Sydowia **9**: 556. 1955.

Stromatic tissues prosenchymatous, forming erumpent stromatic disc; perithecia oblique, circinate, beaks ±lateral, converging and erumpent together through disc. Asci broadly cylindroid or inflated, four or eight spored, paraphyses broad, deliquescent. Ascospores hyaline becoming brown at maturity, ellipsoid oblong, several septate, with vertical septum in one or more of middle cells, rarely in end cell, often appendaged at ends, at times with gel coating, ±uniseriate in the ascus.

Conidial state: not known with certainty (but see *D. acerophila*).

In woody branches.

Type species: *Dictyoporthe ahmadii* Petrak.

Dictyoporthe is quite similar to *Prosthecium* except for the muriform ascospores. The two species previously known in the genus have hyaline ascospores, whereas those of *D. acerophila* are dull brown at maturity.

1. **Dictyoporthe ahmadii** Petrak, Sydowia **9**: 556. 1955.

Stromata circa 1.5 mm diam, tapering to 500-800 μm diam above, disc blackish brown; perithecia 250-350 μm diam or compressed and 140-250 μm diam, up to 500 μm high. Asci 90-120 × 16-23 μm, with elongate stipe 10-20 μm long, eight spored. Ascospores 17-26.5 × 10-12 μm, hyaline, three to five septate and vertical septum in mid cells, uniseriate or partly biseriate in the ascus (from Petrak, **1955b**).

In dead branches of *Spiraea* sp., Pakistan.

Dictyoporthe appendiculata Müller (Sydowia **15**: 87 (1961) 1962) occurs on *Acer opulus* in France and has four-spored asci, the hyaline ascospores (four to six) seven (eight) septate, 31-60 × 13-26 μm, bearing short appendages and surrounded by a gel coating. Müller **(1962)** described ascospore development and recognized similarities with those of *Melanconis appendiculata* (Otth) Wehm. (now included in *Prosthecium*).

2. **Dictyoporthe acerophila** Barr, sp. nov. FIG. 128

Stromata flavofuliginea, erumpentia, perithecia pauciaggregata, globoso-depressa, 240-450 μm diametro, 220-330 μm alto, rostra connivens per discum. Asci 160-250 × 22-25 μm, cylindroidi, unitunicati. Ascosporae 28.5-40 × 13-175 μm, hyalinae vel hepaticae, ellipsoidae, (tribus) quinque (sex) septatae et septum verticalis in cellulis mediis, appendiculatae.

Holotypus in ramis *Aceris sacchari,* "near Pinkham Notch, White Mountains National Forest, New Hampshire, 29 July 1963," a M.E. Barr n. 4007 lecto (MASS).

Stromatic disc light yellow brown, erumpent, perithecia ±circinate, depressed globose, 240-450 μm diam, 220-330 μm high, beaks converging and erumpent through disc. Asci 160-250 × 22-25 μm, broadly cylindroid, apical annulus refractive, nonchitinoid, nonamyloid, eight spored. Ascospores 28.5-40 × 13-17.5 μm, hyaline becoming light dull brown, ellipsoid, (three) five (six) septate, vertical septum in one or two of mid cells, walls and septa thick, appendages hyaline, broad, 2.5-5.5 μm long, 7.5-10 μm wide, uniseriate in the ascus.

A presumed conidial state is associated on the same branches: *Hendersonula*-like: locules formed in stroma; conidiophores short, hyaline; conidia 33-37 × 15-18 μm, brown, obovoid, three to six septate, vertical septum in one to three cells; second conidial type hyaline, lunate, 50-60 × 4-5 μm one to three septate.

Holotype in branches of *Acer saccharum* Marsh., near Pinkham Notch, White Mountains National Forest, New Hampshire, 29 July 1963, Barr 4007 (MASS).

This fungus is similar in many respects to that described by Munk (**1957**) as "*Pseudovalsa* cfr. *umbonata*" on *Acer pseudoplatanus* L. Munk's fungus had longer, transversely septate ascospores; the associated *Coryneum* conidia measured 40-50 × 19-22 μm. *Dictyoporthe acerophila* differs from *Prosthecium stylosporum* (Ell. & Ev.) Wehm., on *Acer spicatum* Lam., in ascospore width and septation, in color of the ectostromatic disc, and in conidial state.

Pseudovalsa macrosperma Tul. var. *fenestrata* Flag. & Chen. (in Chenantais, Bull. Soc. Mycol. France **35**: 124. 1919) was described from *Carpinus* branches. Evidently this taxon belongs in *Dictyoporthe*, as the ascospores were described as brown, 30-32 × 16-18 μm, irregularly muriform, with three transverse and one vertical septum, and short appendaged. Wehmeyer (**1941b**) mentioned this variety in his discussion of *Prosthecium ellipsosporum* and excluded it from *Prosthecium*.

IV-2. Subfamily **Massariovalsoideae** Barr, subfam. nov.

Perithecia aggregata et rostra connata, disco stromate exiguo et pseudoparenchymatico, stromata contexta prosenchymata, vel circumscriptiones per zonas nigrifactos marginalis. (2 tribes). Genus typicus: *Massariovalsa* Sacc.

IV-2a. Tribe **Massariovalseae**

Perithecia grouped, beaks fused together to form a pseudostromatic disc, prosenchymatous tissues slight; in branches of trees and shrubs.

Key to Genera of Tribe Massariovalseae

1. Ascospores one septate, not appendaged, surrounded by gel coating .... 51. *Massariovalsa*
1. Ascospores several septate, appendages elongate, straplike .... 52. *Hapalocystis*

51. **Massariovalsa** Sacc., Michelia **2**: 569. 1882; Syll. Fung. **2**: Addit. LV. 1883.

Melanconis subgenus *Massariovalsa* Wehm., A Revision of *Melanconis*... 48. 1941.

Stromatic tissues prosenchymatous, rather sparse around fused beaks and perithecial walls; perithecia oblique or horizontal, circinate, beaks lateral, converging, fused and forming blackened stromatic disc. Asci cylindroid. Ascospores brown, large, broadly ellipsoid, one septate, walls pigmented, surrounded by gel coating, ± uniseriate in the ascus.

Conidial state: *Melanconiopsis* Ell. & Ev.: locules formed in prosenchymatous stroma; conidiophores annellidic; conidia dark brown, one celled, obovoid or subgloboid, surrounded by gel coating.

In branches of deciduous trees.

Type species: *Massariovalsa sudans* (Berk. & Curt.) Saccardo.

Wehmeyer (**1939, 1941b**) considered that *Massariovalsa* fitted in his broad concept of *Melanconis* as a subgenus. Petrak (**1952b**) described *M. sudans* as a massarioid fungus, similar to *Massariella curreyi* (Tul.)

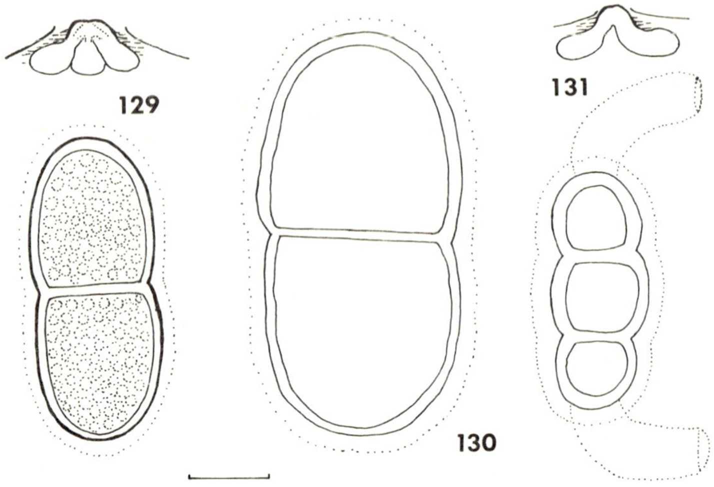

FIGS. 129-131. FIGS. 129, 130. Species of *Massariovalsa*. 129. *M. sudans*, habit of perithecia, ascospore. 130. *M. californica*, ascospore. FIG. 131. *Hapalocystis berkeleyi*, habit of perithecia, ascospore. Scale =1 mm or 10 μm.

Sacc. (*Splanchnonema*) but distinct in valsoid stromata and united elongate ostioles. Müller in Müller and von Arx (**1962**) recognized that *Massariovalsa* possessed unitunicate asci and retained the genus in the Diaporthaceae. Müller also transferred *M. megalospora* (Auersw.) Müller into the genus. This species differs from *M. sudans* in the type of stromatic tissues, centrum, and ascospore pigmentation; it is the type species of *Dothivalsaria* Petrak (Sydowia **19**: 283. 1966 (1965)). Petrak (**1966b**) suggested that *Massariovalsa* as well as *Dothivalsaria* were "dothideaceous." The unitunicate asci of *Massariovalsa sudans* as well as other features exclude this possibility. *Dothivalsaria* appears to be a member of the Massariaceae (see excluded genera).

1. **Massariovalsa sudans** (Berk. & Curt.) Sacc., Michelia **2**: 569. 1882.
FIG. 129

Massaria sudans Berk. & Curt., Grevillea **4**: 156. 1876.

Massariella sudans (Berk. & Curt.) Sacc., Syll. Fung. **1**: 717. 1882.

Melanconis sudans (Berk. & Curt.) Wehm., A Revision of *Melanconis* ... 49. 1941.

Massaria australis Cooke, Grevillea **8**: 67. 1879.

Massariella australis (Cke.) Sacc., Syll. Fung. **1**: 716. 1882.

Melanconis nyssagena Ell. & Ev., J. Mycol. **8**: 18. 1902.

Melanconiella nyssagena (Ell. & Ev.) Sacc., Syll. Fung. **17**: 691. 1905.

Stromatic tissues slight around beaks, brown, disc of fused beaks blackened at substrate surface, interior composed of pallid or yellowish brown hyphae in gel matrix, ostiolar canals converging; perithecia 400-600 μm diam. Asci 265-285 × 22-26.5 μm. Ascospores 30-45(-53) × (11-)15.5-18.5(-23) μm, dark brown, one septate, contents guttulate, surrounded by gel coating 2-3 μm wide.

Conidial state: *Melanconiopsis inquinans* Ell. & Ev.: conidia 28.5-35 × 14-20(-30) μm.

In branches of various deciduous trees, species of *Acer*, *Nyssa*, *Populus*, *Ulmus*, North America.

Material examined: *New York:* New York Fungi 154, 344 (as *Valsaria apatosa* Cooke & Ellis, MASS); *Michigan:* Middleville, Aug 1976, J.R. French (MASS); *New Jersey:* Ellis N.A.F. 1190 (MASS); *Pennsylvania:* Ellis & Everhart N.A.F. 1580 (MASS); Georgia: GA 6721-6734 (GA).

2. **Massariovalsa californica** (Wehm.) Barr, comb. nov. FIG. 130

Melanconis californica Wehm. A Revision of *Melanconis* ... 50. 1941 (as "*Melanconis californica* (Ell. & Harkn.) Sp. nov." for *Melanconis californica* Ellis & Harkness ined.).

Stromatic tissues brown, surrounding perithecia, disc of fused beak tissue, blackened at surface, hyaline internally; perithecia 330-440 μm diam. Asci 220-360 × 30-40 μm. Ascospores 43-70 × 25-30 μm, dark reddish brown, one septate, surrounded by gel coating 4-7 μm wide.

Associated conidial state: *Melanconiopsis:* conidia 30-52(-73) × 22-36 μm.

In branches of *Arbutus* sp., California.

Material examined: *California:* winter 1882, Harkness (NY, type).

52. **Hapalocystis** Auersw. ex Fuckel, Fungi Rhen. no. 585 (in sched.) 1863.

Prosthecium subgenus *Pseudoprosthecium* Wehm., Revision of *Melanconis* ... 102. 1941.

Stromatic tissues prosenchymatous, sparse or well developed, forming a small brown disc around fused ostioles and thin weft of hyphae around perithecia; perithecia oblique or horizontal, circinate; beaks converging and fused together, ostiolar canals converging and united. Asci broadly oblong or ellipsoid. Ascospores brown, at times hyaline until germination, ellipsoid, several septate, occasionally with a vertical septum in one mid cell, wall at times surrounded by gel coating, appendages broad, elongate, straplike.

Conidial state: *Stilbospora*-like: conidial locules shallow in stromatic tissues; conidiophores short, annelidic; conidia hyaline or brown, ellipsoid or cylindroid, several septate.

In branches of deciduous trees.

Lectotype: *Hapalocystis berkeleyi* Auersw. ex Fuckel (Holm, **1975b**).

Holm (**1975b**) observed that *Hapalocystis* Auerswald is a valid and available generic name, although it has not been used since Fuckel (**1870**) relegated the species to other genera. Of the two species described originally, Holm proposed *H. berkeleyi* Auersw. ex Fuckel as lectotype. The removal of *H. berkeleyi* to *Calospora* as *C. hapalocystis* and of the second species *H. bicaudata* to *Melanconis* as *M. berkeleyi* by Fuckel (**1870**) effectively buried the generic name for many years. Wehmeyer's *Prosthecium* subgenus *Pseudoprosthecium* was typified by *P. hapalocystis* (Berk. & Br.) Wehm. (*Hapalocystis berkeleyi*).

The species of *Hapalocystis* differ from those of *Prosthecium* in fused beaks which form a reduced stromatic disc, and in the type of appendages of the ascospores.

1. **Hapalocystis berkeleyi** Auerswald ex Fuckel, Fungi Rhen. No. 585 (in sched.) 1863. FIG. 131

Sphaeria hapalocystis Berk. & Br., Ann. Mag. Nat. Hist. ser. 2, **9**: 317. 1852.

Calospora hapalocystis (Berk. & Br.) Fckl., Symb. Mycol. 191. 1870.

Valsa hapalocystis (Berk. & Br.) Peck, Ann. Rep. New York State Mus. **25**: 104. 1873.

Pseudovalsa hapalocystis (Berk. & Br.) Sacc., Michelia **1**: 44. 1877.

Aglaospora hapalocystis (Berk. & Br.) Kuntze, Rev. Gen. Pl. **3**(2): 441. 1898.

Prosthecium hapalocystis (Berk. & Br.) Petrak, Ann. Mycol. **21**: 324. 1923.

Stromatic tissues forming thin weft of light brown hyphae around fused beaks and upper part of perithecia, perithecia 400-500 μm diam. Asci 75-112 × 30-45 μm, two, four or eight spored. Ascospores 29.5-37.5 (-44) × 14.5-22 μm, (one) two (three) septate, (rarely with a vertical septum in mid cell), gel coating brown, appendages 11-20 μm long, 5.5-9 μm wide, tips chitinoid.

In branches of *Platanus* spp., Europe, North America.

Material examined: *Massachusetts:* Newton, 1893, Farlow; Barr 2871 (MASS); *New York:* Ellis & Everhart N.A.F. 2119 (MASS); *Delaware:* Newark, Feb 1908, H.S. Jackson (GA 6762); *District of Columbia:* Ellis & Everhart Fungi Col. 1843 (MASS).

Hapalocystis berkeleyi var. **kickxii** (West.) Barr, comb. nov. (basionym: *Hypoxylon kickxii* West., Bull. Acad. Roy. Belg. ser. 2, **19**: 112. 1852) has four-spored asci and five-septate ascospores, 55-72 × 22-27 μm. Another species on *Platanus*, **Hapalocystis corticale** (Schw.) Barr, comb. nov. (basionym: *Sphaeria corticale* Schw., Syn. fung. Car. n. 514. 1832) differs from *H. berkeleyi* in more strongly developed stromatic disc surrounding the fused beaks, and larger, three septate ascospores, (35-)51-70 × 13-20 μm, with appendages 9-24 μm long, 4.5-6 μm wide.

Hapalocystis ulmi (Wehm.) Barr, comb. nov. (basionym: *Pseudovalsa ulmi* Wehm., Pap. Mich. Acad. **23**: 237. 1938) is a North American species in which the ascospores do not become brown until germination; conidia are also hyaline. *Hapalocystis ulmi* is similar in many respects to *H. bicaudata* Fuckel (*Prosthecium berkeleyi* (Tul.) Wehm.), on *Ulmus* in Europe. In the European species the ascospores are brown, 38-55 × 15-18 μm, and the conidia (*Stilbospora macrosperma* Berk. & Br.) are also brown, while the hyaline ascospores of *H. ulmi* measure 30-45 × 12-15 μm. Wehmeyer (**1941b**) remarked on the similarities between the two species.

IV-2b. Tribe **Hercosporae** Barr, tribus nov.

Perithecia aggregata, stromata circumscriptiones per zonas nigrifactos marginalis; in ramorum arboribus et frutibus. Genus typicus: *Hercospora* Fries.

Perithecia grouped, beaks fused as pseudostromatic disc, prosenchymatous tissues delimited by blackened marginal zones.

One genus: *Hercospora* Fries.

53. **Hercospora** Fries, Syst. Orbis Veget. 119. 1825; emend. Tulasne & Tulasne, Sel. Fung. Carp. **2**: 154. 1863.

Caudospora Starb., K. Svenska Vetensk. Akad. Handl. **15**: 11. 1889.

Stromatic tissues prosenchymatous around perithecia, delimited externally by blackened dense pseudoparenchymatous zone, interior whitish, composed of interwoven hyphae mixed with substrate cells; perithecia few, small, circinate, beaks converging, becoming united and erumpent through stroma surface as single large opening. Asci broadly cylindroid. Ascospores hyaline, broadly ellipsoid, one septate, wall smooth, without gel coating, with narrow terminal and median appendages in some species.

Conidial state: *Rabenhorstia* Fr.: stromata similar in general aspect to perithecial stromata, pycnidial locule single, opening by ostiole surrounded by a superficial cap of sterile tissue; conidiophores elongate; conidia hyaline, ovoid to ellipsoid, one celled. (Petrak, **1938**; Ruhland, **1900**).

In branches of deciduous trees.

Type species: *Hercospora tiliae* (Pers. ex Fr.) Fries.

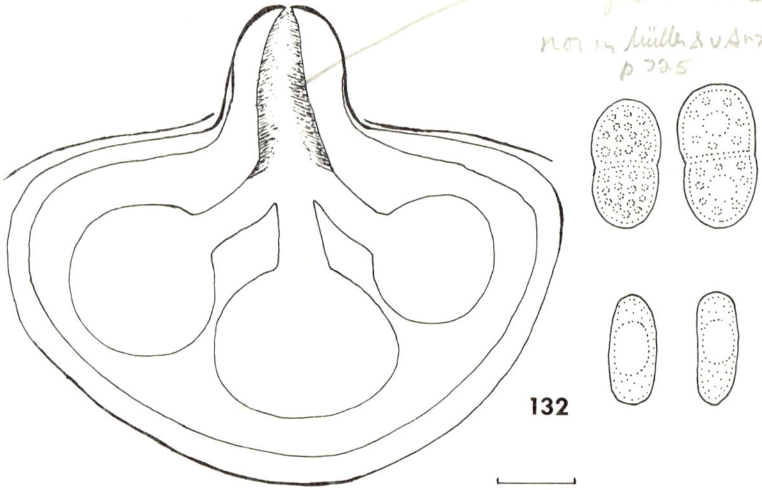

132

FIG. 132. *Hercospora tiliae,* outline of stroma with perithecia, ascospores, conidia. Scale =100 μm or 10 μm.

The typification of *Hercospora* Fries is perhaps questionable. Fries specified that *Sphaeria tiliae* and *S. atrovirens* were two of the species in *Hercospora*, but the generic diagnosis describes ascospores as opaque, while those of *H. tiliae* are hyaline. Tulasne and Tulasne (**1863**) accepted *Hercospora* for *H. tiliae*, thus effectively emending the generic concept. Certainly the majority of mycologists have accepted *H. tiliae* as the type species, while *Sphaeria atrovirens* apparently is synonymous with *Phoma candollei* (Berk. & Br.) Sacc.

Tulasne and Tulasne (**1863**) and Saccardo (**1882**) implicated *Exosporium tiliae* Link. in the life cycle of *Hercospora tiliae*, as well as *Rabenhorstia tiliae*. The fungus which they illustrated as *E. tiliae* is *Corynespora olivacea* (Wallr.) M.B. Ellis, which is common on species of *Tilia* in Europe and North America. *Exosporium tiliae* Link ex Schlecht. is a different species, evidently not occurring in North America. The two were separated and illustrated by Hughes (**1960**) who considered *C. olivacea* as synonymous with *Helminthosporium oligosporum* (Corda) Hughes.

1. **Hercospora tilliae** (Pers. ex Fr.) Fries, Syst. Orbis Veget. p. 119. 1825. FIG. 132

Sphaeria tiliae Pers. ex Fries, Syst. Mycol. **2**: 485-486. 1823.

Spilobolus tiliae (Pers. ex Fr.) Link, Handbuch ... 380. 1833.

Valsaria tiliae (Pers. ex Fr.) de Not., Sfer. Ital. p. 58. 1863.

Sphaeria leprosa Pers. ex Fries, Syst. Mycol. **2**: 365. 1823.

Valsa leprosa (Pers. ex Fr.) Kickx, Rech. Flor. Crypt. Fland. **3**: 16. 1847.

Wuestneia monodelpha Auers. ex Fuckel, Fungi rhen. 594. 1857.

Valsa tilaginea Currey, Phil. Trans. Roy. Soc. London 546. 1857.

Stromata 700-800 μm diam; perithecia 200-250 μm diam. Asci 120-140 × 13-18 μm. Ascospores 15.5-18.5 × 9-11 μm, without appendages.

Conidial state: *Rabenhorstia tiliae* Fries: Conidia 14-16.5 × 4.5-6.5 μm.

In branches of *Tilia* spp., Europe, North America.

Material examined: *Maine:* Barr 5846b; *New Hamphire:* Barr 3879, 4066; *Massachusetts:* Barr 5291, 6279, 6381 (MASS); *Georgia:* GA 6712, 6713 (GA).

Hercospora taleola (Fr.) Müller (in Müller and von Arx, Beitr. Kryptogamenfl. Schweiz **11**(2): 728. 1962) is the type species of *Caudospora* Starb. This species possesses narrow appendages at the tips of ascospores and also two or three arising at the septum. Wehmeyer (**1933**) included the species in *Diaporthe*, where it is alien not only by the ascospore appendages and the sharply delimited, isolated stroma, but also by cylindroid asci. The species is well known in Europe and from North America in Iowa (Gilman, Tiffany and Lewis, **1959**) and North Carolina (GA 6543).

Müller in Müller and von Arx (**1962**) included *Ceratoporthe* Petrak (Ann. Mycol. **23**: 14. 1925) as well as *Caudospora* in the synonymy of *Hercospora*. The type species, *Ceratoporthe didymospora* Petrak (*Hercospora digitifera* (Mout.) Müller), on *Sarothamnus* in Europe, has perithecia which are not closely grouped in the stroma, beaks often erumpent separately, broad asci and narrower ascospores than in typical representatives of the genus. I am doubtful of its position.

HOST INDEX

Many species in the Diaporthales exhibit host specificity, and are limited in occurrence to one or several species of a genus or to closely related genera of host plants. When the host index was compiled, it became apparent that six families in five orders of dicotyledonous plants (Cronquist, **1968**) were hosts to the majority of the species considered. These families are: Juglandaceae in the Juglandales, Fagaceae and Betulaceae (incl. Corylaceae) in the Fagales, Salicaceae in the Salicales, Rosaceae in the Rosales, and Aceraceae in the Sapindales. This information may reflect only the availability of substrate material to collectors, but it is interesting to speculate whether the predominantly woody plants of these families are more susceptible to infection by members of the Diaporthales than are herbaceous plants.

For identification of collections, a host index is useful in reducing the number of possible names. In the following index, where the type species of a genus is not known from temperate North America, the species is included but is indicated by an asterisk. Extralimital species of a genus, however, are not otherwise inserted in the host index. Host families and genera within families are arranged alphabetically. Those species described in the text are listed below, except for a few with wide host ranges, e.g. *Diaporthe eres* and *Valsa ambiens.*

Abiaceae /Pinaceae 163

Aceraceae

Acer

Cryptodiaporthe petiolophila
C. acerinum
Cryptosporella hypodermia
Diaporthella sphendamnina
Dictyoporthe acerophila
Endothia gyrosa
Gnomonia cerastis
G. petiolorum
Macrodiaporthe everhartii
Massariovalsa sudans
Plagiostoma acerophila
P. bavarica
Pleuroceras tenella
Prosthecium stylosporum
Sphaerognomonia carpinea

Anacardiaceae

Rhus

Amphiporthe aculeans
Gnomonia rhuicola

Araliaceae

Aralia

Gnomonia similisetacea

Betulaceae (incl. Corylaceae)

Alnus

Apiognomonia alniella
*Ditopella ditopa
Ditopellopsis alni
Gnomonia alni

G. intermedia var. alni
G. setacea
Gnomoniella tubaeformis
Ophiovalsa suffusa
O. femoralis
Phragmoporthe conformis
Plagiostoma alnea
Pseudovalselle thelebola
Sphaerognomonia carpinea

Betula

Anisogramma virgultorum
Chapeckia nigrospora
Diaporthella aristata
Gnomonia intermedia
G. setacea
Plagiostoma alnea var. betulina
P. campylostyla
P. campylostyla var. mirabilis
Prosthecium acrocystis
Pseudovalsa lanciformis
Sillia spinifera
Xenotypa aterrima

Carpinus

Maniania fimbriata
*Prosthecium ellipsosporum
Sphaerognomonia carpinea

Corylus

Allantoporthe decedens
Anisogramma anomala
Gnomonia amoena
G. gnomon
Mamianiella coryli
M. coryli var. spiralis
Sillia ferruginea
Sphaerognomonia carpinea

Ostrya

Gnomonia ostryae
Mamiania fimbriata
Sphaerognomonia carpinea

Caprifoliaceae

Linnaea

Gnomonia linnaeae

Sambucus

Cryptodiaporthe vepris
Pseudovalsa lanciformis form

Triosteum

Gnomonia triostei

Celastraceae

Euonymus

Gnomonia similisetacea

Cistaceae

Hudsonia

Plagistoma conradii

Clethraceae

Clethra

Ditopellopsis clethrae

Compositae

indet. genera

Diaporthopsis trinucleata
D. apiculosa

Aster

Clypeoporthella appendiculata
Diaporthopsis brencklei

Solidago

Mazzantia biennis

Convolvulaceae

Convolvulus

Diaporthopsis sepium

Cornaceae

Cornus

Cryptodiaporthe corni
Gnomonia rigniacensis

Ericaceae

Arbutus

Massariovalsa californica

Cassiope

Gnomoniella hyparctica

Vaccinium

Gnomonia similisetacea

Gnomonia artospora
Sphaerognomonia carpinea

Onagraceae

Epilobium

Ditopellopsis racemula
Gnomonia comari
Sydowiella fenestrans

Oenothera

Gnomonia misella

Palmae

Sabal

Linocarpon palmetto
L. versisporum

Pandanaceae

Pandanus
*Linocarpon pandani

Platanaceae

Platanus

Apiognomonia veneta
Hapalocystis berkeleyi
H. corticale

Poaceae

indet. genus

*Clypeoporthe monocarpa

Andropogon

Clypeoporthe andropogonis

Oryza

Gaeumannomyces graminis

Saccharum

Clypeoporthe iliau

Polygonaceae

Coccoloba

Gnomonia pulcherrima

Polygonum

Plagiophiale petrakii
Plagiostoma devexa

Polypodiaceae

Onoclea

Gnomonia artospora

Primulaceae

Lysimachia

Clypeoporthella kriegeriana

Rosaceae

Agrimonia

Gnomonia comari

Dalibarda

Gnomonia dalibardae

Dryas

Gnomonia dryadis
G. sibbaldiae
Gnomoniella vagans

Fragaria

Gnomonia comari

Geum

Gnomonia gei-montani

Potentilla

Gnomonia comari
Plagiostoma lugubre
P. tormentillae

Prunus

Pseudovalsa lanciformis

Rosa

Gnomonia rauii
Valseutypella tristicha

Rubus

Cryptodiaporthe vepris
Gnomonia comari
G. rubi
Sydowiella depressula

Spiraea

Cryptodiaporthe macounii
Dictyoporthe ahmadii
Sphaerogonomonia papillostoma

GENERA EXCLUDED FROM THE DIAPORTHALES

A number of genera previously referred to the Diaporthaceae or the Diaporthales belong elswhere in the Ascomycetes according to my classification. These are listed alphabetically.

Acanthonitschkea Speg., Anal. Mus. Nac. Buenos Aires **3**(10): 116. 1908. Type species: *A. argentinensis* Speg. According to von Höhnel (**1917**) this was a species of *Valsa.* Fitzpatrick (**1923**) included the genus in the Nitschkeae (Coronophorales). Nannfeldt (**1975b**) recognized *Acanthonitschkea* as a genus of the Nitschkiaceae in the Sordariales.

Amphisphaeria Ces. & de Not., Comment. Soc. Crittog. Ital. **1**: 223. 1863. Lectotype species: *A. umbrina* (Fr.) de Not. Dennis (**1960**) included the genus in his admittedly heterogeneous assemblage of genera in the Diaporthaceae. *Amphisphaeria* is usually referred to the Amphisphaeriaceae, Xylariales.

Apiorhynchostoma Petrak, Ann. Mycol. **21**: 185. 1923. Type species: *A. curreyi* (Rab.) E. Müller in Müller and von Arx, **1962.** This genus was also placed by Dennis (**1960**) in the Diaporthaceae and is better disposed in the Amphisphaeriaceae.

Apiospora Sacc., Atti Soc. Venet.-Trent. Sci. Nat. **4**: 85. 1875. Type species: *A. montagnei* Sacc. This genus, placed in the Diaporthaceae by Dennis (**1960**), was inserted in the Amphisphaeriaceae by Müller and von Arx (**1962**). Barr (**1976b**) placed it in the Physosporellaceae of the Phyllachorales.

Ceratosphaeria Niessl, Verh. Naturforsch. Ver. Brunn **14**: 208. 1876. Type species: *C. lampadophora* (Berk. & Br.) Niessl. This genus and *Endoxyla* and *Rhamphoria* have been included in the Diaporthaceae (von Arx and Müller, **1954**; Müller and von Arx, **1962, 1973**), the Rhamphorioideae of Diaporthaceae (Munk, **1953**, **1957**), or Ceratostomaceae (Luttrell, **1951**; Chadefaud, **1960**; Dennis, **1960** for *Ceratosphaeria* and *Endoxyla*). The three genera possess some of the characters found in the Diaporthales, i.e., pseudoparenchymatous centrum, asci with chitinoid apical annulus, loosening at times and lying free in the centrum. However, the carbonaceous-walled perithecia and their habit — occurring as secondary saprobes on decorticated wood — are not characteristic of taxa in the Diaporthales. They appear better placed in the Sordariales, for the present in the family Lasiosphaeriaceae.

Ceriophora von Höhnel, Sitzungsber. Kaiserl. Akad. Wiss., Math.-Naturwiss. Cl., Abt. 1, **128**: 586. 1919. Type species: *C. palustris* (Berk. & Br.) von Höhnel. Placed in the Diaporthaceae by Dennis (**1960**), this genus belongs in the Amphisphaeriaceae (Müller and von Arx, **1962**).

Clypeosphaeria Fuckel, Jahrb. Nass. Vereins Naturkunde **23-24**: 117. 1870. Type species: *C. mamillana* (Fr.) Lambotte. Though Dennis (**1960**) inserted *Clypeosphaeria* in the Diaporthaceae, he noted that it appeared instead to be related to *Anthostomella*. It is included in the Xylariaceae by Miller (**1949**), Munk (**1957**), and seems best placed in that family despite the septate ascospores.

Cryptonectriopsis (von Höhnel) von Höhnel, Ann. Mycol. **17**: 131. 1919. (As *Hyponectria* subg. *Cryptonectriopsis* von Höhnel, Ann. Mycol. **16**: 36. 1918.) Type species: *C. biparasitica* von Höhnel. The type species was based upon Rehm's Ascom. exs. n. 1523 as *Phomatospora ovalis* (Pass.) Sacc. According to von Höhnel (**1918**), Passerini's *Sphaeria ovalis* is a pycnidial fungus, *Mycosticta ovalis* (Pass.) von Höhnel, parasitic in the pycnidial state of *Heteropatella lacera*. Rehm's specimen, however, is a species of *Phomatospora*, hyperparasitic in old ascocarps of *Leptosphaeria dolioloides*. Von Arx and Müller (**1954**) included this species in the synonymy of *Heteropera borealis* (Sacc.) Theissen (see entry under *Heteropera*).

Debaryella von Höhnel, Ann. Mycol. **2**: 274. 1904. Type species: *D. hyalina* von Höhnel. Munk (**1957**) referred the fungus to the Diaporthales. The hyperparasitic *D. hyalina* is not diaporthaceous, and although Munk expanded the concept of the genus, the species which he added does not appear to be diaporthaceous either. Perhaps *Debaryella* belongs in the Trichosphaeriaceae of the Xylariales.

Deshpandiella Kamat & Ullasa in Ullasa and Rao, Bull. Torrey Bot. Club **100**: 41-42. 1973. Type species: *D. jambolana* (Ramakr. et al.) Kamat & Ullasa. This was designated as a genus of the Diaporthaceae. The type species causes leaf spotting of *Eugenia cuminii* (L.) Druce (*Syzygium jambolanum* (Lamk.) DC.) in the Myrtaceae. No stroma was described and the lateral perithecia are indistinctly clypeate. Ascospores are one septate, brown. *Mycohyphallage congesta* (Berk. & Br.) Sutton is closely associated and was suggested to be the conidial state (Ullasa and Rao, **1973**). From the accounts in the literature, *D. jambolana* seems to belong in the Amphisphaeriaceae.

Diapleella Munk, Dansk Bot. Arkiv **15**(2): 74. 1953. Type species: *D. clivensis* (Berk. & Br.) Munk. Munk in **1957** placed the genus in the Diaporthoideae of the Diaporthaceae. Dennis (**1968**) removed the genus to the Amphisphaeriaceae, where it seems better placed.

Dothivalsaria Petrak, Sydowia **19**: 283. (1965) 1966. Type species: *D. megalospora* (Auersw.) Petrak. This segregate from *Massariovalsa* (Müller and von Arx, **1962**) is evidently a loculoascomycete. According to configuration of ascocarps and stromatic tissues, pseudoparaphyses, asci and ascospores, *D. megalospora* is most closely related to *Aglaospora profusa* (Fr.) Petrak, in the Massariaceae of the Melanommatales.

Enchnoa Fries, Summa Veget. Scand.: 393. 1849. Lectotype species: *E. infernalis* (Kunze ex Fr.) Fckl. (Clements and Shear, **1931**). The genus has been included in the Diaporthaceae (Munk, **1957**), as well as a member of the Diatrypaceae or of the Coronophoraceae by several authors. I believe that the genus belongs in the Calosphaeriaceae (Barr in Bigelow and Barr, **1969**).

Endoxyla Fuckel, Jahrb. Nass. Vereins f. Naturkunde **25-26**: 321. 1871. Lectotype species: *E. operculata* (Alb. & Schw. ex Fr.) Sacc. (according to von Arx and Müller, **1954**). Synonymy includes *Ceratostomella* Sacc., Michelia **1**: 370. 1878; *Ceratostomis* Clements in Clements and Shear, Genera of fungi, 259. 1931. *Eutypopsis* Karsten, Medd. Soc. Fenn. **2**: 182. 1878; Hedwigia **18**: 22. 1879; *Lentomitella* von Höhnel, Ann. Mycol. **3**: 552. 1905 (von Arx and Müller, **1954**; Müller and von Arx, **1962**). See notes under *Ceratosphaeria*. However, *Endoxyla parallela* (Fr.) Fckl., the type species of *Eutypopsis*, according to specimens of Ellis's N.A.F. 194, is scarcely separable from species of *Cryptosphaeria* Grev. in the Diatrypaceae.

Fenestella Tulasne & Tulasne, Selecta Fung. Carpologia **2**: 208. 1863. Type species: *F. princeps* Tul. & Tul. =*F. fenestrata* (Berk. & Br.) Schroet. This genus was included in the Valsaceae by earlier authors, e.g., Winter, 1887. The type species is a loculoascomycete, belonging in the Melanommatales.

Geminispora Pat., Bull. Soc. Mycol. France **9**: 151. 1893. Type species:

G. mimosae Pat. Lindau in Engler and Prantl (**1897**) included it as a doubtful genus of the Gnomoniaceae. Von Arx and Müller (**1954**) disposed of *Geminispora* as a synonym of *Phyllachora* as Theissen and Sydow (**1915**) had suggested. Von Arx and Müller noted that *Diplosporis* Clements (**1909**) was an unnecessary name given to *Geminispora.*

Glomerella von Schrenk & Spaulding, USDA Bureau Plant Industry Bull. **44**: 29. 1903. Type species: *G. cingulata* (Stoneman) von Schrenk & Spaulding. The genus was proposed to replace *Gnomoniopsis* Stoneman, non Berlese. *Glomerella* was still included in the Diaporthaceae by Wehmeyer (**1975**), although von Arx and Müller (**1954**) placed the genus in the Polystigmataceae. It appears to be related to members of both the Hyponectriaceae and Melogrammataceae (Phyllachoraceae, Polystigmataceae); I believe it is best accommodated in the Melogrammataceae of the Phyllachorales.

Griphosphaeria von Höhnel, Ann. Mycol. **16**: 87. 1918. Type species: *G. corticola* (Fckl.) von Höhnel. This was placed in the Diaporthaceae by Dennis (**1960**). It belongs in the Amphisphaeriaceae (Müller and von Arx, **1973**).

Heteropera Theissen, Ann. Mycol. **14**: 423. 1916. Type species: *H. borealis* (Sacc.) Theissen. The genus was based upon the combination of characteristics of a species of *Cryptonectriopsis* (see that entry) parasitic in ascocarps of a pleosporaceous fungus, as Petrak (**1960**) showed and the name *Heteropera* is invalid.

Kalmusia Niessl, Verh. Nat. Verein Brünn **10**: 204. 1872. Type species: *K. ebuli* Niessl. *Kalmusia* and its subgenus *Thyridaria* Sacc. were arranged in the Valsaceae according to Lindau in Engler and Prantl (**1897**). From my study of available specimens of *K. ebuli,* the fungus is a loculoascomycete and a member of the Melanommatales, family Melanommataceae, related to *Karstenula* Speg.

Lentomita Niessl, Notizen über Pyrenomyceten, p. 44. 1876. Lectotype species: *L. brevicollis* Niessl. Munk included the genus in the Rhamphorioideae of the Diaporthaceae (**1953, 1957**). Müller in Müller and von Arx (**1962**) transferred *L. brevicollis* to *Chaetosphaeria* in the Sphaeriaceae and reduced the genus to synonymy. A second species, *L. caespitosa* Niessl, was transferred to *Otthia* in the Pleosporales by Müller in Müller and von Arx (**1962**).

Magnaporthe Krause & Webster, Mycologia **64**: 110. 1972. Type species: *M. salvinii* (Catt.) Krause & Webster. This fungus deviates especially in the conidial state from members of the Diaporthales. The genus was transferred to the Physosporellaceae of the Phyllachorales (Barr, **1977**), and was enlarged by the addition of *Ceratosphaeria grisea* Hebert (**1971**).

Massariella Speg., Ann. Soc. Cient. Argentina **4**(9): 192 (table) (p. 2 in reprint). 1880. Type species: *M. bufonia* (Berk. & Br.) Speg. Dennis (**1960**) assigned the genus to the Diaporthaceae. Müller and von Arx (**1962**) included *Massariella* as a synonym under *Amphisphaeria* in the Amphisphaeriaceae.

Melanochaeta Müller, Harr & Sulmont, Rev. Mycol. **33**: 377. 1969. Type species: *M. hemipsila* (Berk. & Br.) Müller, Harr & Sulmont. This genus was recently described as a member of the Diaporthaceae. In my opinion it shows closer affinities to members of the Trichosphaeriaceae. The superficial, setose perithecia with papillate apex and the narrow paraphyses suggest *Litschaueria*. The ascospores of *M. hemipsila*, although they are similar to those of *Pseudovalsa*, are basically like those of *Litschaueria*. Conidia borne from phialides are found in several genera of the Trichosphaeriaceae, so that the *Sporochisma* conidial state is not out of place.

Melogramma Fries, Summa Veget. Scand. p. 386. 1849, emend Tulasne & Tulasne, Selecta Fung. Carp. **2**: 84. 1863. Type species: *M. bulliardi* Tulasne & Tulasne =*M. vagans* de Not. Based upon the characteristics of the type species, namely, soft, ±fleshy vinaceous brown, pseudoparenchymatous stroma, perithecia immersed beneath surface and opening by a short periphysate canal, asci clavate, unitunicate, paraphyses rather narrow, ascospores hyaline becoming reddish brown, long fusoid, three septate, the proper position for the genus appears to be in the Phyllachorales, family Melogrammataceae (Barr, **1976b**). The family name is the earliest for the group of genera usually placed in the Phyllachoraceae of Polystigmataceae. The other common species, *Melogramma spiniferum*, differs markedly from *M. vagans* and is a diaporthaceous fungus in the genus *Pseudovalsa*.

Muelleromyces Kamat & Anahosur in Anahosur, Experientia **24**: 849. 1968. Type species: *M. indicus* Kamat & Anahosur =*M. variisporus* (Died.) Ullasa. Ullasa (**1971**) inserted the genus in the Diaporthaceae. The type species is parasitic and forms blackened tarspots of leaves of *Eugenia jambolana* (Myrtaceae). Ascospores are one septate, brown at the tips and hyaline in the middle. The conidial state is *Kamatella longipedicellata* (T.S. & K. Ramakr.) Ullasa, according to Ullasa (**1971**). I think the genus is better disposed in the Amphisphaeriaceae of the Xylariales.

Myrmaeciella Lindau in Engler and Prantl, Nat. Pflanzenfam. **1**(1): 478. 1897. Type species: *M. endoleucum* (Sacc.) Lindau. The genus was erected to replace *Myrmaecium* Sacc. non Nitschke and was assigned to the Melogrammataceae by Lindau (**1897**). Müller in Müller and von Arx (**1962**) reduced *Myrmaeciella* to synonymy and transferred *M. endoleucum* to *Hypocreopsis* in the Hypocreales.

Myrmaecium Nitschke in Fuckel, Jahrb. Nass. Vereins Naturkunde **23-24**: 227. 1870. Type species: *M. rubricosum* (Fr.) Fckl. Lindau in Engler und Prantl (**1897**) placed the genus in the Melogrammataceae. Saccardo (**1880**) had recognized its similarity to *Valsaria* and Müller and von Arx (**1962**) synonymized *Myrmaecium* with *Valsaria*. In **1962**, Müller and von Arx included *Valsaria* in the Diatrypaceae; in **1973** in the Sphaeriaceae.

Neoflageoletia Reid & Booth, Canad. J. Bot. **44**: 450. 1966. Type species: *N. bambusina* (Sydow) Reid & Booth. The genus was included in the Diaporthales by Wehmeyer (**1975**). Reid and Booth (**1966**), however, considered that *Neoflageoletia* was a member of the Polystigmataceae (Melogrammataceae of the Phyllachorales according to Barr, **1976b**).

Neozimmermannia Koorders, Verh. Akad. Wetench. Amsterdam, II, **13**: 59. 1907. Type species: *N. elasticae* (Zimm.) Koorders. Von Höhnel (**1917**) considered the genus as synonymous with *Valsa*. Von Arx and Müller (**1954**) found no difference between *N. elasticae* and *Glomerella cingulata* (Stoneman) von Schrenk & Spaulding.

Ophioceras Sacc., Syll. Fung. **2**: 358. 1883. Lectotype species (von Höhnel, **1909b**): *O. dolichostomum* (Berk. & Curt.) Sacc. Wehmeyer's (**1975**) inclusion of *Ophioceras* in the Gnomoniaceae constitutes the only recent mention of the genus which I have located. The type species, originally collected in Cuba, was recently found in Florida. *Ophioceras* is considered to be a genus in the Lasiosphaeriaceae of the Sordariales (Conway and Barr, 1977). Mycologia 70: 895

Ophiodothella (P. Henn.) von Höhnel, Sitzungsber. Kaiserl. Akad. Wiss., Math.-Naturwiss. Cl., Abt. 1, **119**: 940. 1910. Type species: *O. atromaculans* (P. Henn.) von Höhnel. Miller (**1949**) arranged the genus among the Diaporthaceae, although Boyd (**1934**) had suggested it belonged in the Clypeosphaeriaceae in the Sphaeriales. Müller and von Arx (**1973**) have it in the Polystigmataceae, where I believe the genus is correctly placed (Melogrammataceae, Barr **1976b**). The majority of species in the genus have a well-developed, stromatic clypeus, beneath which several perithecia occur, but one series of species has the clypeus reduced to a small area surrounding one or two perithecial beaks.

Paradidymella Petrak, Ann. Mycol. **25**: 237. 1927. Type species: *P. tosta* (Berk. & Br.) Petrak. Munk (**1953**) and Dennis (**1960**) assigned the genus to the Diaporthaceae. Müller and von Arx (**1962**) reduced *Paradidymella* to synonymy with *Lejosphaerella*, but Müller and Shoemaker (**1965**) referred *P. tosta* to *Clathridium*. *Clathridium* Berlese non Saccardo is *Discostroma*. Whether in this genus, or maintained in *Paradidymella*, *P. tosta* belongs in the Amphisphaeriaceae.

Peroneutypa Berlese, Icones Fung. **3**: 80. 1902. Type species: *P. heteracantha* (Sacc.) Berl. The genus was included in von Höhnel's (**1917**) concept of theValseen. The disposition in Diatrypaceae by Dennis (**1960**) and Wehmeyer (**1975**) seems to be correct. Tiffany and Gilman (**1965**) included *Eutypa heteracantha* Sacc. in the synonymy of *Eutypa saeparia* (Schw.) Tiffany & Gilman.

Petrakiella Sydow, Ann. Mycol. **22**: 230. 1924. Type species: *P. insignis* Sydow. Kobayashi (**1970**) accepted the genus as a member of the Diaporthaceae. The diagnosis suggests the Phyllachorales.

Phomatospora Sacc., Nuov. Giorn. Bot. Ital. **7**: 306. 1875. Type species: *P. berkeleyi* Sacc. This genus is generally considered diaporthaceous. The stipitate asci, narrow paraphyses, and small one-celled ascospores are characteristic of the Amphisphaeriaceae.

Phomatosporopsis Petrak, Ann. Mycol. **23**: 39. 1925. Type species: *P. angelicae* (Fckl.) Petrak. The genus was segregated from *Phomatospora* because of one-septate ascospores. Von Arx and Müller (**1954**) considered it to be a synonym of *Phomatospora*.

Pleomelogramma Speg., Anal. Mus. Nac. Buenos Aires, ser. 3, **12**: 389. 1909. Type species: *P. argentinense* Speg. The genus was described as a dictyosporous *Melogramma* and was included in the Diaporthaceae by Wehmeyer (**1975**). Petrak and Sydow (**1934**) enlarged and emended the generic description and stated that it was related to *Cucurbitaria*, as *Mycosphaerella* is to *Didymella*. The description of the type species is of a fungus close to *Berlesiella nigerrima* (Blox.) Sacc.; it is surely a loculoascomycete.

Pseudodiaporthe Speg., Anal. Mus. Buenos Aires **19**: 359. 1909. Type species: *P. coffeae* Speg. Wehmeyer (**1938**) doubtfully included this in the Diaporthaceae. Bose (**1961**) and Müller and von Arx (**1962**) considered *Pseudodiaporthe* as a synonym of *Massarina* in the Loculoascomycetes.

Pseudomassaria Jacz., Bull. Herb. Boissier **2**: 663. 1894. Type species:

P. chcndrospora (Ces.) Jacz. This genus has been assigned to the Diaporthaceae in the past. Müller and von Arx (**1962**) considered it to belong in the Amphisphaeriaceae, as did Barr (**1964**). I separated *Pseudomassaria* and some other genera from the Amphisphaeriaceae ss. lat., and grouped them in the Hyponectriaceae of the Phyllachorales (Barr, **1976b**).

Rhamphoria Niessl, Verh. Naturforsch. Ver. Brunn **14**: 209. 1876. Type species: *R. delicatula* Niessl. See notes under *Ceratosphaeria.*

Scoptria Nits., Pyren. Germ. 83. 1867. Type species: *S. isariphora* Nits. Von Höhnel included this genus in his **1917** concept of the Valseen, and discussed it in **1918**. It belongs in the Diatrypaceae.

Stigmatopsis Trav., Flora Ital. Crypt. **2**: 213. 1906. Type species: *S. baccarinii* Trav. (p. 214). Müller and von Arx (**1962**) placed the genus in the synonymy of *Diaporthe*. Traverso described it as having a genuine effused stroma, as a *Diatrype* with hyaline two-celled ascospores. From the description and illustration, the fungus appears to be diatrypaceous; in that case *Stigmatopsis* could be a synonym of *Cryptosphaeria.*

Urosporella Atkinson, Bull. Agric. Exp. Sta. Cornell Univ. **3**: 99. 1897. Type species: *U. americana* Atk. =*U. alabamiensis* (Ell. & Ev.) Barr. Synonym: *Urospora* Fabre non Areschoug. This genus was included in the Diaporthaceae by von Höhnel (**1917**); it is better accommodated in the Amphisphaeriaceae (von Arx and Müller, **1954**; Barr, **1966**).

Valsaria Ces. & de Not., Comment. Soc. Critt. Ital. **1**: 205. 1863. Type species: *V. insitiva* (Fr.) Ces. & de Not. Some authors have considered *Valsaria* to be diaporthaceous, others to be pleosporaceous. The type species and a few others belong in the Amphisphaeriaceae in my opinion. Müller and von Arx (**1973**) have it in the Sphaeriaceae.

Valsonectria Speg., Anal. Sci. Cien. Argentina **12**: 211. 1881. Type species: *V. pulchella* Speg. Von Höhnel (**1909a**) combined *Valsonectria* with *Endothia*, based upon the description of *V. pulchella*. However, Petrak and Sydow (**1936**) studied original material and recognized that the type species was hypocreaceous. It is accepted as a genus of the Hypocreaceae by Müller and von Arx (**1962**) and Rogerson (**1970**). *Valsonectria pulchella* was originally described from Argentina on *Melia azedarach* L. A collection from Georgia (GA 7836) on *Pyricularia pubera* Michx. agrees in all respects with descriptions from the literature and appears to be the first collection known from North America.

Vialaea Sacc., Bull. Soc. Mycol. France **12**: 66. 1896. Type species: *V. insculpta* (Fr.) Sacc. The genus was inserted in the Diaportheen by von Höhnel (**1917**) and was included in the Diaporthaceae by Dennis (**1960**) and Müller and von Arx (**1962**). Schrantz (**1960**) erected a new family to accommodate the genus in the Sphaeriales; Müller and von Arx (**1973**) include it in the Amphisphaeriaceae, where it appears to belong.

Vleugelia Reid & Booth, Canad. J. Bot. **47**: 1057. 1969. Type species: *V. betulina* (Bubak & Vleugel) Reid & Booth. This is a peculiar fungus in asci and ascospores. Reid and Booth (**1969**) could not dispose of the genus within a family, but Müller and von Arx (**1973**) included it in the Diaporthaceae. Because of the clavicipitoid asci and the unusual ascospores, I do not consider *Vleugelia* to belong in the Diaporthales, but cannot suggest a good position for the genus.

Winterina Sacc. emend. Sacc. & Sydow in Sacc., Syll. Fung. **14**: 589. 1899. Type species: *W. tuberculifera* (Ell. & Ev.) Sacc. Synonym: *Winterella* Berlese, Icon. Fung. **1**(3): 94. 1892, non Kuntze, 1891, nec Saccardo, 1899. The genus, placed in the Diaportheen by von Höhnel (**1917**), was included in the Coronophorales by Fitzpatrick (**1923**) and was said to be a synonym of *Calyculosphaeria.* Nannfeldt (**1975a**) included the type species in the synonymy of *Nitschkia grevillii* (Rehm in Starb.) Nannf., which he placed in the Nitschkiaceae of the Sordariales.

Zignoëlla Sacc., Michelia **1**: 346. 1878. The presumed type species *Z. ovoidea* Sacc. was included in the Rhamphorioideae of Diaporthaceae by Munk (**1957**). Dennis (**1968**) considered that the genus belonged in the Sphaeriaceae (the Trichosphaeriaceae of Xylariales, according to Barr, **1976a**).

LITERATURE CITED

Alexopoulos, C. J., and D. Cation. 1948. Stem-end rot of strawberries. Phytopathology **38**: 698-706.

———**, and** ———. 1952. *Gnomonia fragariae* in Michigan. Mycologia **44**: 221-223.

Anderson, P. J. 1914. The morphology and life history of the chestnut blight fungus. PennsylvaniaChestnut Tree Blight Comm. Bull. **7**: 1-44.

Arx, J. A. von. 1951. Ueber die Gattung *Laestadia* und die Gnomoniaceen. Antonie van Leeuwenhoek Ned. Tijdschr. Hyg. **17**: 259-272.

———. 1957. Revision der zu *Gloeosporium* gestellten Pilze. Verh. Kon. Ned. Akad. Wetensch. Afd. Natuurk. Tweede Sect. **51**(3): 1-153.

———**, and E. Müller.** 1954. Die Gattungen der amerosporen Pyrenomyceten. Beitr. Kryptogamenfl. Schweiz **11**(1): 1-434.

———**, and D. L. Olivier.** 1952. The taxonomy of *Ophiobolus graminis* Sacc. Trans. Brit. Mycol. Soc. **35**: 29-33.

Auerswald, B. 1869. Synopsis Pyrenomycetum europaeorum. Heft V and VI *in* W. Gonnermann and L. Rabenhorst, Mycologia Europaea.

Barr, M. E. 1959. Northern Pyrenomycetes I. Canadian eastern arctic. Contrib. Inst. Bot. Univ. Montréal **73**: 1-101.

———. 1961. Northern Pyrenomycetes II. Gaspesian Park. Canad. J. Bot. **39**: 307-325.

———. 1964. The genus *Pseudomassaria* in North America. Mycologia **56**: 841-862.

———. 1966. Observations on *Urosporella.* Mycologia **58**: 699-693.

———. 1970. Some amerosporous Ascomycetes on Ericaceae and Empetraceae. Mycologia **62**: 377-394.

———. 1976a. Perspectives in the Ascomycotina. Mem. New York Bot. Gard. **28**: 1-8.

———. 1976b. *Buergenerula* and the Physosporellaceae. Mycologia **68**: 611-621.

———. 1977. *Magnaporthe, Telimenella,* and *Hyponectria* (Physosporellaceae). Mycologia **69**: 952-966.

Bigelow, H. E., and M. E. Barr. 1969. Contributin to the fungus flora of northeasten North America. V. Rhodora **71**: 786-203.

Bolay, A. 1972. Contribution à la connaissance de *Gnomonia comari* Karsten (syn. *G. fruticola* [Arnaud] Fall). Étude taxonomique, phytopathologique et recherches sur sa croissance in vitro. Ber. Schweiz. Bot. Ges. **81**: 398-482.

Bonar, L. 1965. Studies on some California fungi IV. Mycologia **57**: 379-396.

Bose, S. K. 1961. Studies on *Massarina* Sacc. and related genera. Phytopathol. Z. **41**: 151-213.

Boyd, E. S. 1934. A developmental study of a new species of *Ophiodothella.* Mycologia **26**: 456-468.

Brooks, F. T. 1910. The development of *Gnomonia erythrostoma* Pers. The cherry-leaf-scorch disease. Ann. Bot. (London) **24**: 585-605.

Cash, E. K., and A. M. Waterman. 1957. A new species of *Plagiostoma* associated with a leaf disease of hybrid aspens. Mycologia **49**: 756-760.

Cesati, V., and G. de Notaris. 1863. Schema di classificazione degli Sferiacei Italici aschigeri. Comment. Soc. Crittog. Ital. **1**: 177-240.

Chadefaud, M. 1960. Le végétaux non vasculaires (Cryptogamie). *In* Chadefaud, M., and L. Emberger, Traité de Botanique Systematique. Vol. 1 Masson, Paris. 1016 p.

Clements, F. E. 1909. The genera of fungi. A key to Saccardo's Sylloge Fungorum. H. W. Wilson Co., New York. 227 p.

———, and C L. Shear. 1931. The genera of fungi. H. W. Wilson Co., New York. 496 p.

Cole, J. R. 1933. Liver-spot disease of pecan foliage caused by *Gnomonia caryae pecanae,* nov. var. J. Agric. Res. **47**: 869-881.

———. 1935. *Gnomonia nerviseda,* the perfect stage of the fungus that causes the vein spot disease of pecan foliage. J. Agric. Res. **50**: 91-96.

Conway, K. E., and M. E. Barr. 1977. Classification of *Ophioceras dolichostomum.* Mycotaxon **5**: 376-380.

Cronquist, A. 1968. The evolution and classification of flowering plants. Houghton Mifflin Co., Boston. 396 p.

Défago, G. 1942. Seconde contribution á la connaissance des Valsées von Höhnel. Phytopathol. Z. **14**: 103-147.

Demaree, J. B., and J. R. Cole. 1936. A disporous *Gnomonia* on pecan. Phytopathology **26**: 1025-1029.

Dennis, R. W. G. 1960. British cup fungi and their allies. An introduction to the Ascomycetes. Ray Society, London. 280 p.

———. 1968. British Ascomycetes. Cramer, Lehre, 455 p.

———. 1975. New or interesting British microfungi, III. Kew Bull. **30**: 345-365.

Doguet, G. 1961. Étude du *Melogramma spiniferum* (Wallr.) De Notaris, Pyrénomycète, Ascohyménié, Annellascé, Bituniqué. Rev. Mycol. **25**: 13-37.

Edgerton, C. W. 1913. The stem rot or Hawaiian "iliau" disease of sugar cane. Phytopathology **3**: 93-97.

Ellis, J. B., and B. M. Everhart. 1888. New species of fungi from various localities. J. Mycol. **4**: 62-65.

———, and ———. 1892. The North American Pyrenomycetes. Newfield, New Jersey. 793 p.

Fall, J. 1951. Studies on fungus parasites of strawberry leaves in Ontario. Canad. J. Bot. **29**: 299-315.

Fitzpatrick, H.M. 1923. Monograph of the Nitschkieae. Mycologia **15**: 23-67.

———, H. E. Thomas, and R. S. Kirby. 1922. The *Ophiobolus* causing take-all of wheat. Mycologia **14**: 30-37.

Fresenius, G. 1852. Beiträge zur Mykologie **2**: 39-80. Frankfurt a. M.

Fries, E. M. 1923. Systema Mycologicum. Vol. 2. Lundae. 629 p.

———. 1825. Systema Orbis Vegetabilis. Pars 1. 373 p.

———. 1849. Summa Vegetabilium Scandinaviae. Section posterior, pp. 259-572. Uppsala.

Fuckel, L. 1870. Symbolae Mycologicae. Jahrb. Nassauischen Vereins Naturk. **23/24**: 1-459. (Cramer reprint 1966).

Funk, A. 1975. New microfungi on coastal Douglas fir. Canad. J. Bot. **53**: 2297-2302.

Gäumann, E. 1952. The fungi. Translated by F. L. Wynd. Hafner, New York. 420 p.

———. 1964. Die Pilze. Birkhauser Verlag, Basel und Stuttgart. 541 p.

Gilman, J. C., L. H. Tiffany, and R. M. Lewis. 1959. Iowa Ascomycetes III. Diaporthaceae: Diaportheae. Iowa State Coll. J. Sci. **33**: 325-393.

Hebert, T. T. 1971. The perfect stage of *Pyricularia grisea*. Phytopathology **61**: 83-87.

Höhnel, F. von. 1906. Fragmente zur Mykologie (II. Mitteilung, Nr. 64 bis 91). Sitzungsber. Kaiserl. Akad. Wiss., Math.-Naturwiss. Cl., Abt. 1, **115**: 649-695.

———. 1909a. Fragmente zur Mykologie (IV. Mitteilung, Nr. 407 bis 467). Sitzungsber. Kaiserl. Akad. Wiss., Math.-Naturwiss. Cl., Abt. 1, **118**: 1461-1552.

———. 1909b. Fragmente zur Mykologie (CI. Mitteilung, Nr. 182 bis 288). Sitzungsber. Kaiserl. Akad. Wiss., Math.-Naturwiss. Cl., Abt. 1, **118**: 275-452.

———. 1917. System der Diaportheen. Ber. Deut. Bot. Ges. **35**: 631-638.

———. 1918. Mycologische Fragmente. Ann. Mycol. **16**: 35-174.

———. 1919. Mycologische Fragmente. Ann. Mycol. **17**: 114-133.

Holm, L. 1975a. Taxonomic notes on Ascomycetes VIII. Microfungi on *Cassiope tetragona*. Svensk Bot. Tidskr. **69**: 143-160.

———. 1975b. Nomenclatural notes on Pyrenomycetes. Taxon **24**: 475-488.

Homgren, P. K., and W. Keuken. 1974. Index Herbariorum Part I. The Herbaria of the World. 6th Ed. Regnum Veget. **92**: 1-397.

Hubbes, M. 1960a. Systematische und physiologische Untersuchungen an Valseen auf Weiden. Phytopathol. Z. **39**: 65-93.

———. 1960b. Untersuchungen über die Valsaceengattung *Valseutypella* v. H. Phytopathol. Z. **39**: 389-400.

Hughes, S. J. 1955. Microfungi II. *Thyrostromella* Höhn. and *Thyrostromella* Syd. Canad. J. Bot. **33**: 341-349.

———. 1960. Microfungi V. *Conoplea* Pers. and *Exosporium* Link. Canad. J. Bot. **38**: 659-696.

Iqbal, S. H. 1972. *Gaeumannomyces leptosporus.* A new aquatic ascomycete. Trans. Brit. Mycol. Soc. **58**: 346-348.

Jones, S. G. 1926. The development of the perithecium of *Ophiobolus graminis,* Sacc. Ann. Bot. (London) **40**: 607-629.

Karsten, P. A. 1873. Mycologica Fennica. Pars secunda Pyrenomycetes. Bidrag Kännedom Finlands Natur Folk **23**: 1-250. (Cramer reprint 1966).

Katumoto, K. 1965. On the genus *Bagcheea* in Japan. Bull. Fac. Agric. Yamaguti Univ. **16**: 615-622.

Kern, H. 1955. Taxonomic studies in the genus *Leucostoma.* Pap. Michigan Acad. Sci. **40**: 9-22.

———. 1957. Untersuchungen über die Umgrenzung der Arten in der Ascomycetengattung *Leucostoma.* Phytopathol. Z. **30**: 149-180.

———. 1961. Physiologische und systematische Untersuchungen in der Gattung *Leucostoma.* Phytopathol. Z. **40**: 303-314.

Klebahn, H. 1905. Untersuchungen über einige Fungi imperfecti und die zugehörigen Ascomycetenformen. I u. II. J. Wiss. Bot. **41**: 485-560.

———. 1907. Untersuchungen über einige Fungi imperfecti und die zugehörigen Ascomycetenformen. IV. *Marssonia Juglandis* (Lib.) Sacc. Z. Pflanzenkrankh. **17**: 223-237.

———. 1908. Untersuchungen über einige Fungi imperfecti und die zugehörigen Ascomycetenformen. VII. *Leptothyrium alneum* (Lév.) Sacc. Z. Pflanzenkrankh. **18**: 140-154.

———. 1918. Haupt- und Nebenfruchtformen der Askomyzeten. Borntraeger, Leipzig. 395 p. (Cramer reprint 1968).

Kobayashi, T. 1970. Taxomomic studies of Japanese Diaporthaceae with special reference to their life-histories. Bull. Gov. Forest Exp. Sta. **226**: 1-242.

Kohlmeyer, J. 1972. A revision of the Halosphaeriaceae. Canad. J. Bot. **50**: 1951-1963.

Kreisel, H. 1969. Grundzüge eines natürlichen Systems der Pilze. Cramer, Lehre, 245 p.

Kuntze, O. 1891. Revisio Generum Plantarum. 1. 1011 p. (Pars II Fungi, pp. 840-875).

Likkité, V. 1926. Recherches sur le développment et la biologie de quelques Ascomycètes. Rev. Gen. Bot. **38**: 5-30.

Lind. J. 1928. The Micromycetes of Svalbard. Skr. Svalbard Ishavet **13**: 1-61.

Lindau, G. 1897. Pyrenomycetineae. *In* A. Engler and K. Prantl, Eds. Die natürlichen Pflanzenfamilien **1**(1): 321-491.

Luttrell, E. S. 1951. Taxomomy of the Pyrenomycetes. Univ. Missouri Stud. **24**(3):: 1-120.

Miles, L. E. 1921. Leaf spots of the elm. Bot. Gaz. (Crawfordsville) **71**: 161-196.

Miller, J. H. 1941. Georgia Pyrenomycetes. II. Mycologia **33**: 74-81.

———. 1949. A revision of the classification of the Ascomycetes with special emphasis on the Pyrenomycetes. Mycologia **41**: 99-127.

———**, and G. E. Thompson.** 1940. Georgia Pyrenomycetes I. Mycologia **32**: 1-15.

———**, and F. A. Wolf.** 1936. A leaf-spot disease of honey locust caused by a new species of *Linospora.* Mycologia **28**: 171-180.

Montagne, J. F. C. 1856. Sylloge Generum specierumque cryptogamarum quas in variis operibus descriptas iconibusque illustratas, nunc ad diagnosim reductas, non-nullasque novas interjectas, ordine systematic exposiit. Paris, J. B. Baillière. 498 p.

Morgan-Jones, J. F. 1953. Morpho-cytological studies of the genus *Gnomonia.* I. *Gnomonia intermedia* Rehm: Its development in pure culture. Svensk Bot. Tidskr. **47**: 284-308.

———. 1958. Morpho-cytological studies of the genus *Gnomonia*. II. The asexual stage of *Gnomonia ulmea* (Schw.) Thüm. Svensk. Bot. Tidskr. **52**: 363-372.

———. 1959. Morpho-cytological studies of the genus *Gnomonia.* III. Early stages of perithecial development. Svensk Bot. Tidskr. **53**: 81-101.

Müller, E. 1957. Pilze aus dem Himalaya I. Sydowia **11**: 454-472.

———. (1961) 1962. Über einige Ascomyceten aus Südfrankreich. Sydowia **15**: 84-91.

———. (1964) 1965. Beobachtungen an Ascomyceten. Sydowia **18**: 86-105.

———**, and J. A. von Arx.** 1950. Einige Aspekte zur Systematik pseudosphärialer Ascomyceten. Ber. Schweiz. Bot. Ges. **60**: 329-397.

———, and ———. 1962. Die Gattungen der didymosporen Pyrenomyceten. Beitr. Kryptogamenfl. Schweiz **11**(2): 1-922.

———, and ———. 1973. Pyrenomycetes: Meliolales, Coronophorales, Sphaeriales. Pp. 87-132. *In* G. C. Ainsworth, F. K Sparrow, and A. S. Sussman, Eds. The fungi, Vol. IVA. Academic Press, New York and London.

———, **and R. Menon.** 1954. *Bagcheea*, eine neue Gattung der Sphaeriales aus Indien. Phytopathol. Z. **22**: 417-420.

———, **and R. A. Shoemaker.** 1965. The ascogenous state of *Seimatosporium*(≡*Monoceras*) *kriegerianum* on *Epilobium* species. Canad. J. Bot. **43**: 1343-1345.

Munk, A. 1953. The system of the Pyrenomycetes. Dansk Bot. Arkiv. **15**(2): 1-163.

———. 1957. Danish Pyrenomycetes A preliminary flora. Ejnar Munksgaard, Copenhagen. 491 p.

Nannfeldt, J. A. 1931. Contributions to the mycoflora of Sweden. Svensk Bot. Tidskr. **25**: 1-31.

———. 1932. Studien über die Morphologie und Systematik der nicht-lichenisierten inoperculaten Discomyceten. Nova Acta Regia Soc. Sci. Upsal., ser. 4. **8**(2): 1-368.

———. 1975a. Stray studies in the Coronophorales (Pyrenomycetes) 1-3. Svensk Bot. Tidskr. **69**: 49-66.

———. 1975b. Stray studies in the Coronophorales (Pyrenomycetes) 4-8. Svensk Bot. Tidskr. **69**: 289-335.

Neely, D. 1968. The somatic pressure cushion of *Gnomonia platani.* Mycologia **60**: 84-89.

———, **and E. B. Himelick.** 1965. Nomenclature of the sycamore anthracnose fungus. Mycologia **57**: 834-837.

———, **and** ———. 1967. Characteristics and nomenclature of the oak anthracnose fungus. Phytopathology **57**: 1230-1236.

Negru, A. 1964. Quelques espèces de Micromycetes nouvelles de Romanie. Bull. Soc. Mycol. France **80**: 223-231.

Niessl, G. v. 1875. Notiz zur 20. Centurie der Fungi europ. Hedwigia **14**: 129-130.

Nitschke, T. 1867. Pyrenomycetes Germanici. Breslau. 320 p.

Petrak, F. 1914. Beiträge zur Pilzflora von Mähren und Österr.-Schlesien. Ann. Mycol. **12**: 471-479.

———. 1919. Mykologische Notizen. I. Ann. Mycol. **17**: 59-100.

———. 1921. Mykologische Notizen. II. Ann. Mycol. **19**: 17-128.

———. 1923. Mykologische Notizen. VI. Ann. Mycol. **21**: 182-335.

———. 1927. Mykologische Notizen. IX. Ann. Mycol. **25**: 193-343.

———.1934. Mykologische Notizen. XII. Ann. Mycol. **32**: 317-447.

———. 1938. Beiträge zur Kenntnis der Gattung *Hercospora* mit besonderer Berücksichtigung ihrer Typusart *Hercospora tiliae* (Pers.) Fr. Ann. Mycol. **36**: 44-60.

———. 1940. Mykologische Notizen. XIII. Ann. Mycol. **38**: 181-267.

———. 1941. Mykologische Notizen. XIV. Ann. Mycol. **39**: 251-349.

———. 1950. Fungi beltsvillenses. I. Sydowia **4**: 402-408.

———. 1952a. Fungi beltsvillenses. III. Sydowia **6**: 5-16.

———. 1952b. Über die Gattung *Massariovalsa* Sacc. Sydowia **6**:: 284-287.

———. 1952c. Über die Gattungen *Gaeumannomyces* v. Arx et Olivier, *Halophiobolus* Linder und *Linocarpon* Syd. Sydowia **6**: 383-388.

———. 1954. Ergebnisse einer Revision der Grundtypen verschiedener Gattungen der Ascomyceten und Fungi imperfecti. Sydowia **8**: 287-303.

———. 1955a. *Xenotypa* n. gen., eine neue Diaportheen-Gattung. Sydowia **9**: 497-500.

———. 1955b. *Dictyoporthe* n. gen., eine neue Gattung der Diaportheen. Sydowia **9**: 556-558.

———. 1955c. *Plagiophiale* n. gen., eine neue Gattung der sphaerialen Askomyzeten. Sydowia **9**: 585-587.

———. 1960. Ergebnisse einer Revision der Grundtypen verschiedener Gattungen der Askomyzeten und Fungi imperfecti. Sydowia **14**: 347-354.

———. (1964) 1965. Über die Gattungen *Plagiostigme* Syd., *Bioporthe* Petr. und *Plagiophiale* Petr. Sydowia **18**: 380-387.

———. (1965) 1966a. Über die Gattung *Cryptospora* Tul. Sydowia **19**: 268-278.

———. (1965) 1966b. Über *Valsaria megalospora* Auersw. und die Gattung *Massariovalsa* Sacc. Sydowia **19**: 279-281.

———. (1969) 1970. Ergebnisse einer Revision der Grundtypen verschiedener Gattung der Ascomyceten und Fungi imperfecti. Sydowia **23**: 265-272.

———. (1970) 1971a. Ergebnisse einer Revision der Grundtypen verschiedener Gattung der Ascomyceten und Fungi imperfecti. Sydowia **24**: 249-255.

———. (1970) 1971b. Über *Diaporthe hranicensis* Petr. Sydowia **24**: 256-260. 261-263.

———. (1970) 1971d. Über die Gattung *Skottsbergiella* Petr. Sydowia **24**: 264-266.

———. (1970) 1971e. Kritische Bemerkungen zur Nomenklatur der Gattung *Discula* Sacc. Sydowia **24**: 270-273.

———, and H. Sydow. 1923. Kritisch-systematische Originaluntersuchungen über Pyrenomyzeten, Sphaeropsideen und Melanconieen. Ann. Mycol. **21**: 349-384.

———, and ———. 1934. Kritisch-systematische Originaluntersuchunge über Pyrenomyzeten, Sphaeropsideen und Melanconieen. Ann. Mycol. **32**: 1-35.

———, and ———. 1936. Kritisch-systematische Originaluntersuchungen über Pyrenomyzeten, Sphaeropsideen und Melanconieen. Ann. Mycol. **34**: 11-52.

Pirozynski, K. A. 1974. *Xenotypa* Petrak und *Graphostroma* gen. nov., segregates from Diatrypaceae. Canad. J. Bot. **52**: 2129-2135.

———, and G. Morgan-Jones. 1968. Notes on microfungi. III. Trans. Brit. Mycol. Soc. **51**: 185-206.

Pomerleau, R. 1938. Recherches sur le *Gnomonia ulmea* (Schw.) Thüm. Contrib. Inst. Bot. Univ. Montréal **31**: 1-139.

Potebnia, A. 1910. Beiträge zur Micromycetenflora Mittel Russlands. Ann. Mycol. **8**: 42-93.

Rehm, H. 1906. Zum Studium der Pyrenomyceten Deutschlands, Deutsch-Österreichs und der Schweiz. Ann. Mycol. **4**: 471-482.

Reid, J., and C. Booth. 1966. The genus *Flageoletia* (Sacc.) Höhn. Canad. J. Bot. **44**: 445-451.

———, and ———. 1967. The Diaporthaceae I. *Ditopellopsis* gen. nov., *Ditopellina* gen. nov., *Phragmoporthe* and *Ditopella*. Canad. J. Bot. **45**: 1479-1487.

———, and ———. 1969. Some species segregated from the genera *Cryptospora, Cryptosporella,* and *Sillia.* Canad. J. Bot. **47**: 1055-1060.

———, **and R. F. Cain.** 1960. Additional Diaporthales on twigs from Ontario. Canad. J. Bot. **38**: 945-950.

Riess, H.1854. Neue Kernpilze. Hedwigia **1**: 25-28.

Roane, M. K., R. J. Stipes, F. M. Phipps, and O. K. Miller, Jr. 1974. *Endothia gyrosa,* causal pathogen of pin oak blight. Mycologia **66:** 1042-1047.

Rogerson, C. T. 1953. Kansas mycological notes: 1951. Trans. Kansas Acad. Sci. **56**: 53-57.

———. 1970. The hypocrealean fungi (Ascomycetes, Hypocreales). Mycologia **62**: 865-910.

Ruhland, W. 1900. Untersuchungen zu einer Morphologie der stromabildenden Sphaeriales auf entwicklungsgeschichtlicher Grundlage. Hedwigia **39**: 1-79.

Saccardo, P. A. 1875. Conspectus generum pyrenomycetum italicorum additis speciebus fungorum Venetorum novis vel criticis, systemate carpologico dispositorum. Atti Soc. Venet.-Trent. Sci. Nat. Padova **4**: 77-100.

———. 1880. Fungorum extra-Europaeorum. Michelia **2**: 136-149.

———. 1882. Sylloge fungorum, Vol. I. Patavii.

———. 1883. Sylloge fungorum, Vol. II. Patavii.

———. 1905. Sylloge fungorum, Vol. XVII. Patavii.

———, **C. H. Peck, and W. Trelease.** 1904. The fungi of Alaska. Harriman Alaska Expedition **5**: 11-64.

Schrantz, J. P. 1960. Recherches sur les Pyrénomycètes de l'ordre des Diatrypales, sensu M. Chadefaud, 1957. Bull. Soc. Mycol. France **76**: 305-407.

Schroeter, J. 1893. Die Pilze Schlesiens. *In* Kryptogamen-Flora von Schlesien **3**(2): 1-256. (**1894**: 257-384; **1897**: 385-500; **1908**: 501-597). (Cramer reprint 1973).

Shear, C. L. 1943. Mycological notes. VII. Mycologia **35**: 469-476.

———, **N. E. Stevens, and H. F. Bain.** 1931. Fungous diseases of the cultivated cranberry. USDA Techn. Bull. **258**: 1-58.

———, ———, **and R. J. Tiller.** 1917. *Endothia parasitica* and related species. USDA Bull. **380**: 1-82.

Shoemaker, R. A. 1974. Fungi Canadenses. # 37.

———, **and P. M. LeClair.** 1975. Type studies of *Massaria* from the Wehmeyer collection. Canad. J. Bot. **53**: 1568-1598.

Sivanesan, A. 1970. *Plurisperma dalbergiae* gen. et sp. nov. Trans. Brit. Mycol. Soc. **54**: 495-496.

Sutton, B. C. 1975. Coelomycetes. V. *Coryneum.* Mycol. Pap. **138**: 1-224.

———. 1977. Coelomycetes VI. Nomenclature of generic names proposed for Coelomycetes. Mycol. Papl. **141**: 1-253.

Sydow, H. 1936. Novae fungorum species—XXIV. Ann. Mycol. **34**: 411-422.

———, **and P. Sydow.** 1916. Weitere Diagnosen neuer philippinischer Pilze. Ann. Mycol. **14**: 353-375.

Theissen, F. 1916. Beiträge zur Systematik der Ascomyzeten. Ann. Mycol. **14**: 401-439.

———, **and H. Sydow.** 1915. Die Dothideales. Kritischsystematische Originaluntersuchungen. Ann. Mycol. **13**: 149-746. (Cramer reprint 1963).

Thompson, G. E. 1954. The perfect stages of *Marssonina rhabdospora* and *Septogloeum rhopaloideum*. Mycologia **46**: 652-659.

Tiffany, L. H., and J. C. Gilman. 1965. Iowa Ascomycetes IV. Diatrypaceae. Iowa State J. Sci. **40**: 121-161.

Traverso, J. B. 1906. Flora Italica Cryptogama Pars I: Fungi Pyrenomycetae. Fasc. **1**: 1-352.

———. 1907. Flora Italica Cryptogama Pars I: Fungi Pyrenomycetae. Fasc. **2**: 353-492.

———. 1913. Flora Italica Cryptogama Pars I: Fungi Pyrenomycetae. Fasc. **3**: 493-700.

Tulasne, L.-R., and C. Tulasne. 1863. Selecta Fungorum Carpologia, Vol. 2. Paris, 319 p. + 34.pl.

Ullasa, B. A. 1971. Identity and status of two species of *Diplodia* affecting *Syzygium* species. Bull. Torrey Bot. Club **98**: 1-4.

———, **and V. G. Rao.** 1973. *Deshpandiella*, a new genus of the Diaporthaceae (Ascomycetes). Bull. Torrey Bot. Club **100**: 41-43.

———, ———, **and A. S. Patil.** (1973) 1974. *Chadefaudiomyces*, a new member of Diaporthaceae (Ascomycetes). Rev. Mycol. **38**: 19-21.

Walker, J. 1972. Type studies on *Gaeumannomyces graminis* and related fungi. Trans. Brit. Mycol. Soc. **58**: 427-457.

Weese, J. 1935. Eumycetes selecti exsiccati. 30. Lieferung, No. 726-750. Mittheil. Bot. Inst. Techn. Hochsch. Wien **12**: 1-32.

Wehmeyer, L. E. 1923. The imperfect stage of some higher Pyrenomycetes obtained in culture. Pap. Michigan Acad. Sci. **3**: 245-266.

———. 1926. A biologic and phylogenetic study of the stromatic Sphaeriales. Amer. J. Bot. **13**: 575-645.

———. 1933. The genus *Diaporthe* Nitschke and its segregates. Univ. Michigan Stud. Sci. Ser. **9**: 1-349.

———. 1938. Les species de"*Diaporthe*" en el herbario Spegazzini. Rev. Mus. de la Plata, n.s. sect. Bot. **2**: 65-88.

———. 1939. The genus *Massariovalsa*. Amer. J. Bot. **26**: 831-834.

———. 1941a. *Pseudotrichia* and the new genus *Phragmodiaporthe*. Mycologia **33**: 54-63.

———. 1941b. A revision of *Melanconis, Pseudovalsa, Prosthecium,* and *Titania*. Univ. Michigan Stud. Sci. Ser. **14**: 1-161. (Cramer reprint 1974).

———. 1942. Contributions to a study of the fungous flora of Nova Scotia. VI. Pyrenomycetes. Canad. J. Res., C, **20**: 572-594.

———. 1961. A world monograph of the genus *Pleospora* and its segregates. Univ. Michigan Press, Ann Arbor. 451 p.

———. 1975. The pyrenomycetous fungi. Mycol. Mem. **6**: 1-250.

Weste, G. 1972. The process of root infection by *Ophiobolus graminis.* Trans. Brit. Mycol. Soc. **59**: 133-147.

Winter, G. 1887. Pilze: Ascomyceten. *In* Rabenhorst's Kryptogamen-Flora von Deutschland, Oesterreich und der Schweiz. **1**(2): 1-928. (Cramer reprint 1963).

Wolf, F. A. 1912. A new *Gnomonia* on hickory leaves. Ann. Mycol. **10**: 488-491.

TAXONOMIC INDEX

Boldface page numbers indicate pages with descriptions